Insects Did It First

Insects Did It First

Gregory S. Paulson
and Eric R. Eaton

Cover photo:

Problems with insects; a pear psylla nymph with a parasitic wasp emerging from it, colorized scanning electron micrograph.

Copyright © 2018 by Gregory S. Paulson and Eric R. Eaton.

Illustrations By: E. Paul Catts

Library of Congress Control Number:	2018913275
ISBN: Hardcover	978-1-9845-6463-4
Softcover	978-1-9845-6462-7
eBook	978-1-9845-6461-0

All rights reserved. No part of this book may be reproduced or transmitted in any form or by any means, electronic or mechanical, including photocopying, recording, or by any information storage and retrieval system, without permission in writing from the copyright owner.

The views expressed in this work are solely those of the author and do not necessarily reflect the views of the publisher, and the publisher hereby disclaims any responsibility for them.

Any people depicted in stock imagery provided by Getty Images are models, and such images are being used for illustrative purposes only. Certain stock imagery © Getty Images.

Print information available on the last page.

Rev. date: 11/13/2018

To order additional copies of this book, contact:
Xlibris
1-888-795-4274
www.Xlibris.com
Orders@Xlibris.com
787748

Contents

Dedication ix
About the Authors xi
Preface to the Second Edition xiii

Introduction 1
Transportation 4
    Flight 4
    Jet Propulsion 5
    Highways 6
    Bridges 7
    Compass 8
    Following Signs 9
    Hitchhiking 11
    Ballooning 12
    Surfing 13
    References 14
Agriculture 16
    Farming 16
    Domestic Animals 17
    Sowing Seeds 19
    Fungicides 20
    Insect Control 20
    References 22
Communication 23
    Blinking Neon Sign 23
    Good Vibrations 25
    Drummer Boys/Percussion 26
    Fiddling Around 27
    References 28

Architecture and Engineering 30
    Tunnel Builders 30
    Air Conditioning 31
    Food Storage 33
    Stonework 34
    Recycling 35
    Sewing or Lashing 36
    Wallpaper 38
    Construction Materials 38
    Pipe Liners 40
    Reinforced Tubing 41
    Mud and Masonry 42
    Ladders 44
    References 45

Chemistry 47
    Antifreeze 47
    Perfumes and Pheromones 48
    Glue 50
    Glue II 51
    Drug Use 52
    Natural Polyesters (Bouncy Chemistry) 54
    Preservation Without Freezing 55
    Dyes 56
    Sunblock 58
    References 59

Tools 61
    Thermometer 61
    Clocks 63
    Tool Use 64
    SCUBA 65
    Fishing Nets 67
    Woodworking Tools 68
    References 69

Family Life & Society 71
    Social Behaviors/Societies 71
    Dominance Hierarchies 72
    Gift-giving 74
    Caste 76
    Cannibalism 77

Drug Use 78
　　Begging 79
　　Robbers 81
　　Communes 82
　　Child Protective Services 83
　　Baby Food 84
　　Determining the Gender of Your Offspring 86
　　References 86

Arts and Entertainment 89
　　Acrobats 89
　　Bungee Jumpers 90
　　Humming 92
　　Dancing 93
　　Sculpture 94
　　References 95

Department of Defense & Warfare 96
　　Chemical Weapons 96
　　Camouflage 98
　　Warning Colors 99
　　Radar/Sonar 101
　　Radar-jamming Device 102
　　Kamikaze, Suicide Bomber 104
　　Armor 105
　　Mimicry 107
　　Arms Race 108
　　Cannon 109
　　Prisoners of War 111
　　References 113

Miscellaneous Categories 115
　　Vampire 115
　　Polluters 117
　　Gravediggers 118
　　Food Storage 119
　　Hypodermic 121
　　Antibiotics/Medicines 122
　　Recycling 123
　　Problems with Insects 125
　　Opposable Thumb 127
　　The Wheel 128

Mechanical Gears 129
Velcro 130
Baskets and Pots 132
Flypaper 133
GPS 134
References 135

Epilogue 137

Dedication

GSP - I would like to thank my late parents, Mr. and Mrs. Neil A. Paulson, Sr., for their support and encouragement through the years. I'd also like to acknowledge the great influence that Dr. Sally L. Paulson and Mr. Neil A. Paulson, Jr., my sister and brother, had on my life. Where would I be today without the Bee Club? Finally, this is dedicated to Pam.

ERE - Dedicates this book to the two late authors of the original edition, and to his wife, Heidi, his late parents, and mentors past, present, and future.

About the Authors

Gregory S. Paulson - Dr. Paulson's career in entomology has been devoted to the applied side of the science. He is especially interested in developing alternatives to pesticides for insect control. He served as a Peace Corps volunteer in Western Samoa in a WHO filariasis research program and studied plant pathology in Hawaii. Most recently he studied ant population structure in the forests of Pennsylvania. Recently retired, he is a professor emeritus of biology at Shippensburg University (PA) and lives in Oregon with his wife, Pam.

Eric R. Eaton – Mr. Eaton is a writer specializing in natural history. He has worked as an entomologist for the Cincinnati Zoo and on private contract for the Smithsonian Institution, and various state and federal agencies. He is principal author of the *Kaufman Field Guide to Insects of North America*, and writes the blogs *Bug Eric* and *Sense of Misplaced*. He resides in Colorado with his wife, Heidi.

Preface to the Second Edition

The authors of the second edition wish first and foremost to acknowledge two of the prior authors, giants in the world of entomology who are no longer with us. The late Roger D. Akre and E. Paul Catts were both professors of entomology at Washington State University. Dr. Akre was one of the first experts in the emerging discipline of urban entomology and an authority on yellowjackets, carpenter ants, and dangerously venomous spiders, among other arthropods. As an instructor, he was also responsible for turning countless students into future entomologists, as well as creating respect and admiration for the invertebrate world through courses in general entomology, insect behavior, insect photography, and the use of insects in teaching.

Dr. Catts was a pioneer in resurrecting of the lost science of forensic entomology, elevating this field to a degree of prominence and respect that endures to this day and inspiring characters like Gil Grissom in television's original *CSI: Crime Scene Investigation*. Dr. Catts was also a distinguished expert in medical and veterinary entomology. At Washington State University he taught courses in insect morphology, medical entomology, and "Insects and People".

It was Dr. Akre who stumbled upon the idea for the first edition of *Insects Did It First* while he was teaching a general entomology course in 1964. Each time some advanced human technology, such as radar or sonar, became the topic of the day, Roger realized that "people aren't really all that original — insects did it first". Soon he started writing down all the ideas that occurred to him on the topic, even if it meant interrupting his own lectures.

Dr. Akre also started to make color slides with the idea that he could someday illustrate "firsts" as a special topic in his general entomology class. Enter Dr. Catts, Dr. Robert Harwood, and Dr. William B. Garnett of the University of Cincinnati. All three gentlemen were gifted artists who turned some of those "firsts" into cartoons that were used to complement Akre's lectures.

As word spread and interest grew, the thought of producing a book came to mind for Akre in 1986. Shortly thereafter Dr. Paulson located a book with a similar theme, *Nature Thought of It First*, an English translation of the 1971 European title *Wonderbaarlijke Nature*, by Lucy Berman and Roy Combs. This spurred Dr. Akre to seriously entertain the idea of his own book. Ross E. Hutchins, a popular nature writer of that time, produced *Nature Invented It First* in 1980. Neither of these books was exclusively devoted to insects, and Dr. Akre saw his opportunity.

The authors of the second edition are honored to continue the legacy of our predecessors through this encore edition, including Catts' cartoon art, and by incorporating new "firsts" discovered since the inaugural volume. The second edition is organized differently than the first edition. Topics are grouped into units by general theme e.g. Transportation, and references are included at the end of each unit to make it easier for someone to access sources for further reading. Thank you for coming along for the ride.

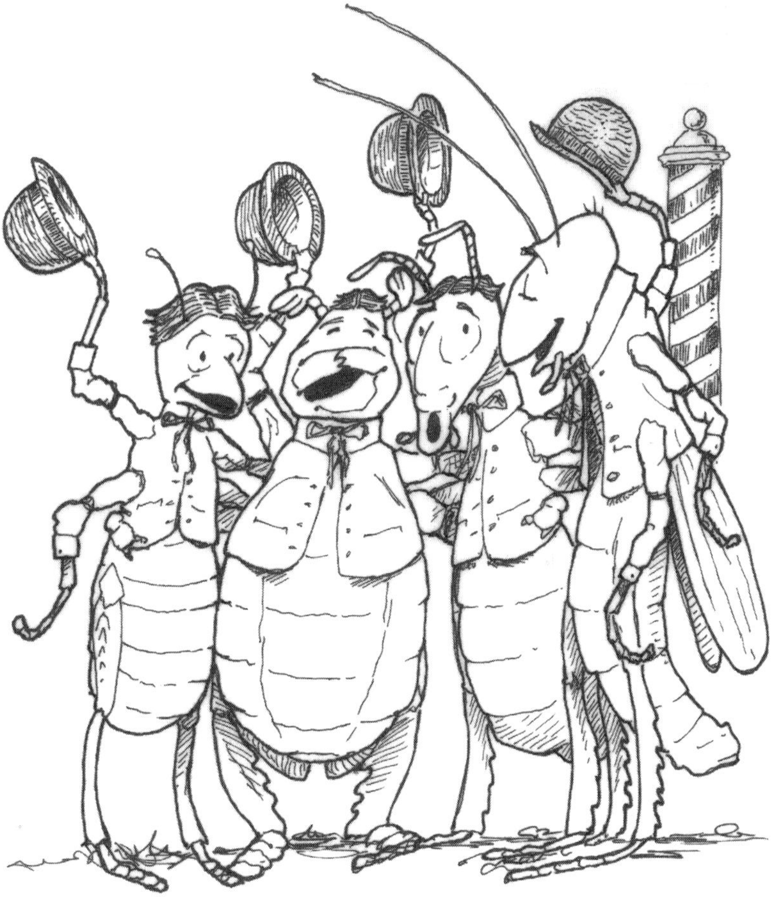

Introduction

The success of insects is almost entirely a function of instinct, diversity, small size, and durability; and their anatomical features are the product of evolution. So how is it that we can credit them with inventing....anything? That is not literally what this book is asserting. The world of insects is often viewed as dull, boring, and not worth the time it takes to swat a mosquito, when in reality "bugs" are mind-blowing examples of organisms vital to the healthy functioning of planet Earth and our own personal and collective well-being.

The idea that insects share behaviors and "inventions" similar, if not identical, to our own human creations and social structure is disturbing to some. But we, too, are animals, and we deny that fact at the risk of our future survival. Why not embrace other life forms with a dedication to understanding them and learning from their success? The chapters that follow aim to help in that endeavor of appreciation.

Insects are indeed successful. No other organisms come close to their sheer diversity and individual numbers. Because of their roles in natural and agricultural ecosystems, *we* have succeeded in feeding ourselves, modeling our own tools and products after theirs, healing ourselves with medicines produced by insects, and creating art and architecture inspired by our six-legged friends. We could not live without insects pollinating crops, controlling pests, recycling decaying plants and animals back into the soil, and producing valuable products like honey, beeswax, silk, natural dyes, and shellac. Aquatic insects are valuable in assessing water quality. Some flies and beetles that feed on carrion can help forensic scientists solve homicides by indicating the time of death. We therefore owe at least some modern inventions or improvements to insects.

Let's look at two case histories. Paper wasps (**Hymenoptera: Vespidae**) scrape wood fibers from fence posts and dead trees and then chew them into a pulp they use to build their paper nest combs. These wasps are credited with inspiring modern paper manufacturing. A shortage of cotton and linen, the raw material standard for paper in sixteenth-century Europe, resulted in a paper crisis. In 1719 French naturalist and physicist Antoine Ferchault de Réaumer, who was inspired by his observations of paper wasps, suggested using wood as a papermaking fiber. The pulp and paper industry continue to rely on wood for paper manufacturing to this day.

The story of how the chainsaw was improved also involves an insect. Until the mid-1940s, saw chains mimicked the tooth arrangement of manual crosscut saws. They were inefficient, dulled quickly, and forced many loggers to reject them in favor of the manual tools. After observing the wood-boring larvae of the Ponderous Borer (**Coleoptera: Cerambycidae**), *Trichocnemis spiculatus*, Joe Cox was inspired to create a chain of alternating "right" and "left" teeth that cut in a manner similar to the insect's opposable mandibles. His design was patented, and he began producing his invention in 1947 in the basement of his Portland, Oregon home under the company name Oregon Sawchain Corporation, which a decade later became Omark Industries as his sawchains entered foreign markets.

But this is the digital age, when technology is advancing almost daily. What could insects still possibly teach us? The answer: a lot more. Technological advances have given us new tools to investigate organisms in novel ways, and the emerging science of biomimicry has focused extensively on insects as models for everything from robotics to social systems. These are exciting times, and if we understand the value of basic research to make the discoveries, then the applied research technologies that can result are almost limitless.

Transportation

Flight

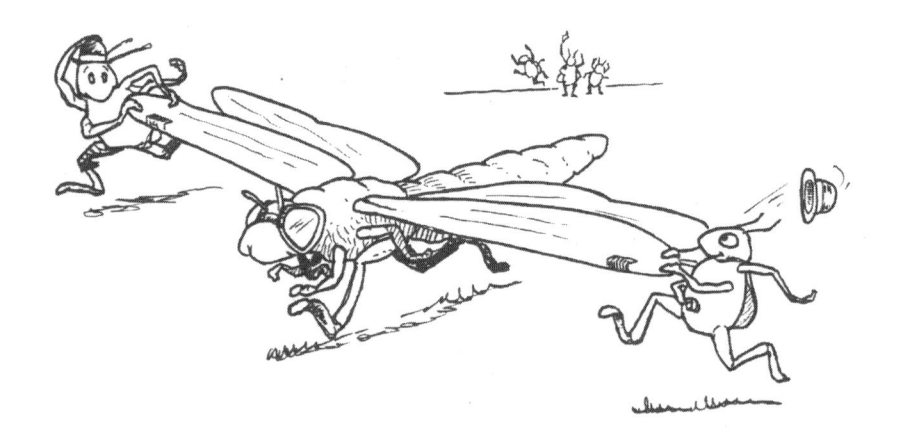

Only birds, bats, insects, and humans are capable of true flight. Other mammals, as well as some reptiles, amphibians, and fish, have mastered gliding. Obviously, the first into the air were the insects, and they are still the most accomplished, agile, and versatile fliers. Most insects have two pairs of wings that they use in tandem; or independently, the forewings and hind wings flapping alternately. True flies (**Diptera**) have only one pair; the hind wings have been reduced to gyroscopic stabilizers called **halteres**. The basic element of insect flight is a click mechanism that suddenly transmits nearly all the stored, latent muscle energy of the thorax to the kinetic energy of the downward, power stroke of the wings.

Variations in flight styles include insects that have the ability to hover in one place. Dragonflies, bot flies, bee flies, flower flies (also known as hover flies), and some bees have perfected hovering. These insects possess a unique flip mechanism that causes the wings to twist and sweep obliquely up and down through a narrow angle. Certain damselflies in the American tropics are known as "helicopter damselflies" because when the insect is hovering, the large, colored spots on either the front pair of wings, or both pairs, are reminiscent of the colored rotors of some helicopters. These damselflies hover in front of spider webs in order to pluck the spider from its resting place. Occasionally they also steal the spider's strung-up meals of trapped insects.

The modern age of robotics and technological engineering still sees humans learning from insect flight. The dynamics of dragonfly flight, for example, are staggering and counter-intuitive. Dragonflies generate turbulence to achieve lift. Turbulence is the arch-enemy of conventional fix-winged human aircraft. Meanwhile, our tiny drones are poor, clumsy facsimiles of their living insect models. Recent advances in our understanding of insect neurology have allowed researchers to "pilot" live beetles in the laboratory, much like in the model aircraft hobby.

Jet Propulsion

Humans are in a constant pursuit of speed, and one of the ways we achieved great velocity early on was the invention of jet propulsion. Still, insects were way ahead of us.

The immature stages of dragonflies and damselflies are aquatic predators called naiads. Despite their fearsome nature, they are pursued by their own set of predators and must therefore be able to escape. Most of the time, dragonfly (Odonata) naiads are well-camouflaged on the bottom of a pond or among reeds and other vegetation close to shore. Should they be discovered, they switch to a different tactic that is the opposite of remaining motionless. Dragonfly naiads are able to forcefully eject water from a special rectal chamber and out the anus. This action can propel them at a speed of up to 4 inches per second for short bursts.

Highways

It might seem ludicrous to draw an analogy between the superhighways of human civilizations and primitive trails made by insects, but the two serve much the same function. In many cases, the trails made by insects are more complex than you might imagine.

Among the most noticeable of all insect trails are those of tropical army ants (**Hymenoptera: Formicidae**). These consist of unbroken lines of ants who are often traveling in both directions. Army ants are blind; they rely completely on chemical (pheromone) trails for navigation. The ants therefore follow trails marked by fellow workers and often reinforce the trail by depositing their own dose of pheromones as they travel. Think of it as an odorous version of GPS, or the olfactory equivalent of traffic signals and other informational cues.

Army ants are also carnivorous and systematically roam their habitat in search of other insects to prey upon. The best-known army ant species are also nomadic and move distances of up to one hundred yards daily before establishing a new bivouac, a temporary nest created by their own, interlocked bodies and appendages.

Many other ants use chemical trails and "highways". The broad trails of leafcutting ants (*Atta* spp.) are an obvious feature in many subtropical and tropical landscapes. These paths which in some stretches are up to 4-6 inches wide, are carefully maintained by workers that clear away all obstructions, including vegetation, so that the trails are bare and well-groomed. Workers use these routes to forage for leaves that they cut into pieces and hold aloft in their jaws as they carry the fragments back to their nest. This habit has earned them the nickname "parasol ants".

Harvester ants in the genus *Pogonomyrmex* gather seeds as well as dead and injured insects for food for their colonies. They often create cleared foraging paths many yards long that radiate from their nest mound.

Even carpenter ants may establish trails similar to those described for leafcutters. In the Pacific Northwest it is not unusual to find trails 1–1.5 inches wide cut through manicured lawns. Since these ants do most of their foraging at night, these trails are physical routes the ants navigate by following pheromone trails.

One thing you never see on an ant freeway is a traffic jam. The insects make travel look effortless. Perhaps we can learn from them how to avoid the snarls and snags and delays that plague our own daily commutes.

Bridges

The problem of crossing rivers or chasms was solved by early humans with the construction of bridges from logs or vines. The art of bridge-building progressed steadily from these times as both an art form and an engineering challenge.

However, among the earliest bridge builders were the ants, particularly the army ants of the Neotropical regions (Central and South America). For example, workers of *Eciton burchelli* form bridges between vines and roots by using their own bodies. The ants simply hold onto each other to form a chain that acts as a living bridge that can then be crossed by the remaining members of the colony to speed their journey. Workers also build living flanges of more ants on vertical and horizontal surfaces of the bridge to widen the pathway.

Compass

Mariners of old who needed to travel in a certain fixed direction relied heavily on celestial bodies, such as the sun and the stars, for direction in the open ocean. Many insects, particularly ants and honey bees, also use the sun for orientation. Recently, it was discovered that certain scarab beetles also orient their dung-rolling direction by heavenly cues. Foraging honey bees (**Hymenoptera: Apidae**) are able to travel at a fixed angle to the sun to return unerringly to the hive. Once there, they convey these directions to nestmates inside the darkened hive. They do this through a "waggle dance" that instructs other worker bees how to navigate to the food source. Incredibly, they even compensate for changes in the sun's position throughout the day. However, the sun cannot be relied upon during heavily overcast weather when it is obscured by clouds. At such times the bees use a compass.

Natural magnets such as lodestone (magnetite or Fe_3O_4) that could be used to point to magnetic north were prized by early human travelers. These primitive compasses evolved into the compasses commonly used in the pre-GPS age. But magnetic navigation systems have been used by other organisms for eons. Even certain bacteria are **magnetotactic** and tend to swim along lines influenced by the Earth's magnetic field. Their compasses consist of magnetite, which they synthesize from soluble iron. More sophisticated uses of the Earth's magnetic field for orientation are used by honey bees. Honey bees can detect, and will orient their dances to, the Earth's magnetism. Their compasses may consist of a region of transversely-oriented material in the front of the **gaster** (abdomen behind the **pedicel**/stalk that joins it to the thorax), as well as bands of cells in each abdominal segment that contain iron granules.

No kind of compass is completely foolproof, and there is little doubt that most insects find themselves frequently pausing and "recalculating", just like we do on long road trips and in unfamiliar neighborhoods. They likely don't stop and ask for directions, either.

Following Signs

Many years ago an old beekeeper from British Columbia said that the entire life of a bee is spent following signs. While it may be simplistic to refer to the behavior of bees in these terms, the statement has a basis in fact. One of the strongest signs or signals used to direct bees are "traffic signs" displayed on flowers; these are more appropriately referred to as "nectar guides". Bees, like the majority of insects, are capable of perceiving the ultraviolet end of the light spectrum. It is why moth collectors use black lights to lure moths to a sheet. Invisible to the human eye, these wavelengths are easily seen by insects. Many flowers have ultraviolet patterns on their petals that guide bees to the center of the flower where the nectar reservoir is located. This is also an area where the bee is most likely to come into contact with the plant's pollen. The flower thus benefits by achieving pollination from the bee's visit, provided the insect has visited other flowers of the same kind previously.

The signs flowers put out are essentially advertisements, but just like modern ads, they can also be deceiving. For example, the shape of some European orchids resemble female wasps in the family **Scoliidae**, and the male wasps are drawn to them thinking they are in for a good time of another kind. Meanwhile, orbweaver spiders (**Araneae: Araneaidae**) in the genus *Argiope* spin a conspicuous zig-zag band of thick silk called a **stabilimentum** near the center of their webs. The silk bands reflect ultraviolet light, and one theory is that this functions as a flower mimic to attract insect prey to the spider's snare.

Insects are more **chemotactile** than visual, however, so the "signs" they are reading are often decoded through touch and especially smell. Flowers capitalize on this, too, of course, and the fragrances wafting from blossoms are meant more for pollinating insects than our romantic and sentimental human enjoyment.

Carnivorous plants like sundews, the Venus flytrap, and pitcher plants rely on scent to draw insect prey to their doom. Scientists are learning that some spiders even lace the sticky silk in their webs with chemicals that mimic the sex pheromones of moths to lure the male moths to death instead of a receptive female.

Hitchhiking

Back in the 1970s, American parents admonished their teenage sons and daughters to not pick up hitchhikers. There is wisdom in that rule for the sake of human safety, but many insects and their relatives have been getting around on the backs of other organisms for a long time.

The term for hitchhiking in the animal kingdom is "phoresy". One extraordinary example is the Human Bot Fly (**Diptera: Oestridae**), *Dermatobia hominis*, of the American tropics. Because the female fly is a large insect that easily attracts the attention and repellent efforts of its human host, she employs another insect to complete her life cycle. She accosts a female mosquito (**Diptera: Culicidae**) in midair and attaches her eggs to the mosquito's body. The mosquito then unwittingly ferries the fly's embryonic offspring to a human host. When the mosquito lands and prepares to bite, the warmth of the human host causes the bot fly eggs to hatch. Each maggot then tunnels into the human's flesh, perhaps using the mosquito's bite wound as a point of entry. Gross, yes, but ingenious.

An insect relative that uses phoresy to move from one location to another is the pseudoscorpion (**Pseudoscorpiones**). These arachnids resemble a tiny scorpion with no tail. Many pseudoscorpion species are associated with trees, where they live under the bark while preying on even smaller organisms. They usually get from tree to tree by crawling onto the back of a longhorned woodborer beetle and slipping beneath the wing covers, where they can ride in relative comfort. They must surely have a strong grip to adhere to the beetle's abdomen once it takes flight. Alternatively, a pseudoscorpion may simply use one of its pincers to grab the leg of any available insect and hold on tight until the living vehicle alights once again.

Some kinds of mites, which are also arachnids, live in flowers and, lacking wings, need help in getting from place to place. Enter hummingbirds. When a hummingbird visits a flower full of mites, the mites swiftly run onto the bird's tongue, bill, and facial feathers.

There are many other species of mites that ride on specific types of insects. Some travel on bees, some on wasps, and some on dung and carrion beetles. Those on the beetles disembark when the beetles arrive at a dung pile or animal carcass. The mites feed on the eggs of flies, and thus pay the "fare" for their ride by eliminating the beetles' competition.

Ballooning

Some of humanity's first flights were in baskets suspended below airtight bags of silk or other fabric filled with lighter-than-air, or heated, gases.

Insects can be balloonists, too. One would think that given the ability of most insects to fly, there would be no need for passive transportation. Surprisingly, the dispersal phase for many moth species is not the adult insect, but rather the caterpillar stage. For example, the newly-hatched caterpillars of the Gypsy Moth (**Lepidoptera: Erbidae**), *Lymantria dispar*, drop from the treetops on strands of silk to be caught and transported by the wind. In addition to silk, the **lateral setae** (hairs) on the bodies of the larvae aid in their buoyancy. This is an effective strategy, as even slight breezes can transport the tiny caterpillars for miles. Females of many of their relatives in the tussock moth family lack wings completely. The females lay their eggs on the cocoons that they emerge from, and the hatchling caterpillars sail away on the wind.

Spiders also use balloons for dispersal. In the spring hundreds of small, newly-hatched spiderlings are transported great distances by this means. Even juvenile spiders and, rarely, adults will utilize this form of transportation to reach new territory. The phenomenon of "gossamer" occurs when the silk strands of thousands upon thousands of ballooning spiders drift to Earth and coat grass and other vegetation in fields. Bon voyage!

Surfing

Wait, insects can surf? Groovy! Hang loose, dude....Human thrill-seekers need boards or other objects, or even motorized transportation to ride the waves, but insects are able to exploit a phenomenon of water known as "surface tension" to achieve water-walking capabilities.

The most familiar occupants of the surface film are water striders, aka "pond skaters", true bugs (**Hemiptera**) in the insect family **Gerridae**. There are also water measurers (**Hydrometridae**), smaller water striders (**Veliidae**), water treaders (**Mesoveliidae**), and velvet water bugs (**Hebridae**) that share similar niches. Sure, these insects are small, and weigh little, but most of them would still sink were it not for extra-long legs that spread their weight across a larger area, much as snowshoes help bulky humans navigate deep snow. The only truly marine insects are the sea striders (**Gerridae**) in the genus *Halobates*. Some are found in the open ocean near the equator. The front legs of water striders are modified to grab and hold prey that is trapped on the water surface. The remaining two pairs of legs are very long and splayed in four directions to support the insect.

Many spiders, especially fishing spiders and wolf spiders, are also able to dash across the surface of the water, but unlike water striders they do not spend the bulk of their time doing so.

The terrestrial rove beetles (**Coleoptera: Staphylinidae**) in the genus *Stenus* can, if necessary, run across the water as well and can enhance the speed at which they do so by secreting a surfactant from a gland in the tip of the abdomen. The effect of the chemical is to drive the beetle forward at impressive speed by lowering the surface tension behind it.

The most astonishing example of surfing insects are ants. In Venezuela, several species of ants have been observed body surfing on small (to us, huge to them) shore breaking waves. Instead of being swept out to sea when struck by a wave the ants assume a characteristic body position and ride the wave to shore, and safety. Unlike humans the ants apparently surf for survival not for fun, but maybe if our hearing was more acute we could hear tiny calls of "kawabunga" as the ants ride to shore. The Red Imported Fire Ant, *Solenopsis invicta*, is well known for surviving floods by forming a living raft of all nest occupants and floating to the nearest dry land. This is doubtless a terrifying prospect to anyone in the southern U.S. who has already gone to great lengths to avoid this stinging pest.

References

Atkins, M.D. 1980. Introduction to insect behavior. MacMillan, New York.

Blakemore, R. P. and R. B. Frankel. 1981. Magnetic navigation in bacteria. Sci. Amer. 245: 58-65.

Bonanos, C. 1992. The father of modern bridges. Amer. Hert. Invent. Tech. 8: 8-20.

Borror, D. J., C. A. Triplehorn, and N. F. Johnson. 1989. An introduction to the study of insects. Saunders, Philadelphia, PA.

Chapman, R. F. 1982. The insects: Structure and function. Harvard University, Cambridge, MA.

Dyer, F. and J. L. Gould. 1983. Honey bee navigation. Amer. Sci. 71: 597.

Hansen, L. D. and R. D. Akre. 1985. Biology of carpenter ants in Washington State (Hymenoptera: Formicidae: *Camponotus*). Melanderia 43: 1-62.

Jaffe, K. 1993. Surfing ants. Florida Entomologist 76: 182-183.

Petroski, H. 1992. The Britannia tubular bridge. Am. Sci. 80: 220-224.

Pringle, J. W. S. 1975. Insect flight. Oxford Biology Reader, No. 52. Oxford Univ. Press, Oxford, England.

Rettenmeyer, C. W. 1963. Behavioral studies of army ants. Univ. Kansas Sci. Bull. 44: 281-465.

Ross, H. H., C.A. Ross, and J. R. P. Ross. 1982. A textbook of entomology. Wiley, New York.

Schneirla, T. C. 1971. Army Ants: A study in social organization. Freeman, San Francisco, CA. von Frisch, K. 1966. The dancing bees: An account of the life and senses of the honey bee. Methuen, London.

von Frisch, K. 1967. The dance language and orientation of bees. Belknap/Harvard Univ. Press, Cambridge, MA.

Wehner, R. 1976. Polarized light navigation by insects. Sci. Amer. 235: 106-115.

Weis-Fogh, T. 1975. Unusual mechanisms for the generation of lift in flying animals. Sci. Amer. 233: 80-87

Agriculture

Farming

The original farmers and gardeners were insects. Leafcutting ants collect leaf fragments that they then compost underground as a medium for growing a type of fungus they use as food for themselves and their brood.

Some termites in Africa and Asia also grow fungus for food. Among these termites (**Blattodea: Termitidae**) are species of *Macrotermes* which have large, air-conditioned nests.

These insect-farmed fungi are able to break down cellulose, a major component of plant tissues. Most insects, including the farming insects, can digest cellulose only with the help of specialized microorganisms, called symbionts, that live in the gut of some insects. Insects without such microbes can still utilize abundant cellulose resources by growing a fungus that does the same job. The fungus is frequently a single species found only in ant or termite nests and kept free of microbial contaminants by secretions from the insect farmers. The fungal mass can consist of small pockets found throughout the nest, or it may be a single large, spongy ball. One such ball found in a termite colony was two feet in diameter and weighed sixty pounds.

Ants and termites can be extremely destructive to cultivated crops and human structures while harvesting material to grow fungus. In Central and South America, leafcutting ants, usually *Atta cephalotes*, may cause extensive defoliation of citrus trees. Equally well known is the destruction of all kinds of crops by termites in Africa and Asia. Some fungus-growing termites will fertilize their fungus crop with their own excrement. Little goes to waste in the animal kingdom, and we could learn from the efficiency of other species.

Domestic Animals

Humanity has domesticated a few animals to use as beasts of burden, as food, and as pets. The more common domestic animals are cows, horses, pigs, chickens, dogs, and cats. However, insects began keeping their own livestock much earlier than we did.

One such association is between certain ants and the larvae of gossamer-winged butterflies, family **Lycaenidae**. The caterpillars of some species possess special glands, called **Hinton's glands**, which produce a substance the ants like to eat. In return, the ants protect the caterpillars from insect predators and parasites.

A more familiar partnership is presented by ants tending aphids and related true bugs such as treehoppers (**Membracidae**), planthoppers (**Fulgoridae**), mealybugs (**Pseudococcidae**), and soft scale insects (**Coccidae**). This relationship is called **mutualism** because both the ants and the aphids profit from the association. **Mutualistic** relationships exist between many genera and species of ants. Not only do ants "herd", or even carry, aphids to their host plants and protect them there, but at least one ant species responds aggressively to an alarm **pheromone** (chemical signal) released by the aphids. When the aphids are disturbed, the ants will attack any insects or other intruders in the immediate vicinity.

Aphids and their kin suck plant sap, which is relatively poor in nutrients, so the insects must take in great quantities of the liquid. As fast as they imbibe the sap, they excrete copious amounts of sugary, liquid waste called honeydew. Ants and many other insects crave this carbohydrate-rich fluid and flock to aphid colonies to drink it. One Middle Eastern species of mealybug, *Trabutina mannipara*, produces so much of this sticky material that it is collected by both ants and people for food. It is believed to be the "manna" mentioned in the Bible.

Sowing Seeds

The legendary figure Johnny Appleseed was a real-life hero named John Chapman who was responsible for creating small orchards across Pennsylvania, Ohio, and Illinois in the early 1800s. At the time, settlers could stake a claim to a parcel of land with an orchard as proof of ownership. Chapman sold his apple seedlings to pioneers, and he died at the age of 70 as the owner of more than 1,200 acres. Insects, too, are seed dispersers, but unlike Chapman, they do not always profit from their labors.

Ants in particular spread the seeds of specific plants. The plants rely heavily on the ants and attract them with a nutrient-rich coating or growth on their seeds. These "**elaiosomes**" or "**arils**" are energetically less expensive for a plant to produce than a fruit that might attract birds or mammals. Oftentimes, the plant cannot afford to produce an elaborate seed vessel due to poor soil nutrients.

Ants carry these seeds back to the nest, where the adult ants and their larvae consume the fat, protein, starch, sugars, and/or vitamins contained in the elaiosomes. The seeds benefit by being concealed and protected from seed-eating animals, and the soil inside an ant nest is often richer for the excrement generated by the occupants. In fire-prone habitats, seeds moved by ants are either safe from fire entirely or may be heated just enough to trigger germination. Some Australian stingless bees (**Hymenoptera: Meliponinae**) harvest resins from eucalyptus trees as building material for their nests. Seeds often adhere to the load of resin a bee carries away from the plant. Back at the nest, or during a pause in its flight, the bee will discard the seeds, thereby dispersing them.

In Europe, tiny seeds of Hooked Foxtail, *Setaria verticillata*, and other grasses in the Poaceae family lodge in the dense hairs of the flower-visiting scarab beetle (**Scarabaeidae**) *Trichius fasciatus* and are transported via the flying insect to destinations far from the original plant.

Dung beetles and termites occasionally benefit plants by burying seeds in their burrows or nests. Who would have thought that sowing seeds is a task accomplished by insects as well as human farmers?

Fungicides

Today, the most frequently applied agricultural chemicals in the United States are herbicides that are used to control weeds in our crops. Some of these are even genetically integrated into the crop plants themselves. Insects also have weed problems. The leafcutting ant, *Atta sexdens*, needs to control "weeds" in the form of unwanted fungi in its food crop, which is itself a fungus grown as food for the colony. These weedy fungi could be a real problem for the ants were it not for a fungicide, **myrmicacin** (beta-hydroxy-decanoic acid), which is produced by the ants in their metathoracic glands. The myrmicacin does not affect the beneficial fungus in their culture, so it can be used to keep their "crop" free of weed fungus.

Insect Control

Insects are unquestionably their own worst enemies. Were it not for the natural parasites or predators of insect pests on crop plants, there would be substantially less food available for human consumption around the world. We are slowly evolving in our pest-control strategies to take advantage of these natural allies. The approach taken nowadays is one of Integrated Pest Management (IPM) that uses chemical controls as a last resort.

Foremost among the predatory insects are ants, which many scientists consider the dominant organisms in most ecosystems, be they natural or artificial. Ants were the first insects used by humans as biological pest-control agents. Weaver ants, genus *Oecophylla*, were used to control citrus pests in China more than a thousand years ago. Additional insect predators include many types of beetles, true bugs like assassin and ambush bugs, and flies like robber flies, wasps, dragonflies, lacewings, and antlions.

In addition to predators, there are many species of parasitic insects that kill their hosts. These are correctly called parasitoids to distinguish them from parasites that are not normally fatal to their hosts. There are many species of parasitoids, but most are tiny wasps or flies.

While all these examples concern insect control as they benefit mankind, the insects themselves also benefit from these interactions, which are essential to maintaining an ecological balance. The well-being of herbivores, predators, parasites, and their shared habitat are all intertwined in the complex fabric of life.

References

Batra, S. W. T. and L. R. Batra. 1967. The fungus gardens of insects. Sci. Amer. 217: 112-120.

Kistner, D. H. 1982. The social insects' bestiary, pp.1-244. *In* H. R. Herman [ed.], Social insects, vol. 3. Academic, New York.

Schildknecht, H. 1971. Evolutionary peaks in the defensive chemistry of insects. Endeavour 30: 136-141.

Vigni, I. and M. R. Melati. 1999. Examples of seed dispersal by entomochory. Acta Bot. Gallica 146 (2): 145-156.

Weber, N. A. 1972. Gardening ants, the Attines. Amer. Phil. Soc., Philadelphia, PA.

Wheeler, W. M. 1973. The fungus-growing ants of North America. Dover, New York.

Wilson, E. O. 1971. The insect societies. Belknap/Harvard Univ. Press, Cambridge, MA.

Wiltz, B. A., G. Henderson, and J. Chen. 1998. Effect of napthalene, butylated hydroxytoluene, dioctyl phthalate, and adipic dioctyl ester, chemicals found in the nests of the Formosan subterranean termite (Isoptera: Rhinoternitidae) on a saprophytic *Mucor* sp. (Zygomycetes: Mucorales). Environ. Entomol. 27(4): 936-940.

Communication

Blinking Neon Sign

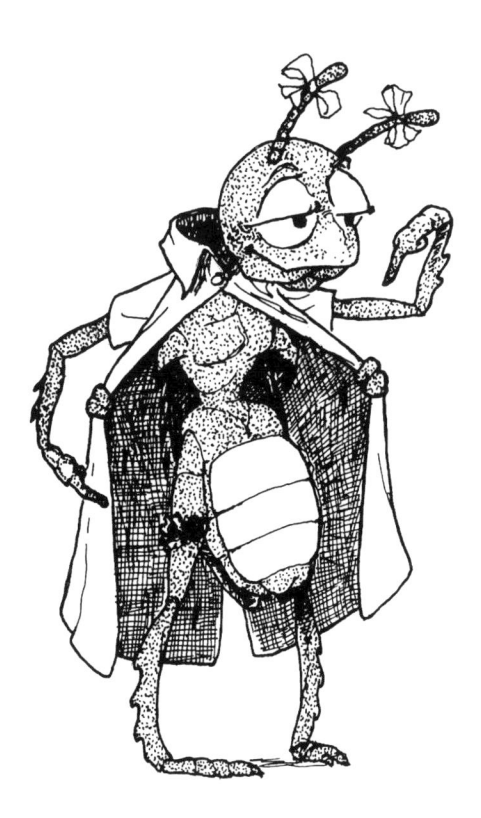

Flashing neon lights are surefire attention-getters that have been used in advertisements for years, but they are slowly being replaced with light-emitting diodes (LEDs) in the interest of energy conservation. Indeed, as early as 1976, during the gasoline shortage, power companies threatened to shut off the lights of the gambling casinos in Las Vegas and Reno as a means of saving oil and countless megawatts of electricity.

But long before people used flashing lights as beacons, fireflies and glowworms employed light-generating organs to both alert potential predators and to attract mates. Fireflies are beetles in the family **Lampyridae**, and glowworms are in the sister family **Phengodidae**. While the larvae of all known firefly species produce light, not all the adult insects do. It is suspected that the light displays evolved first to warn predators of the toxic nature of fireflies, much like the aposematic "warning colors" that day-active insects use to advertise their distastefulness. Fireflies are full of potent poisons related to the skin toxins of toads.

The adults of many firefly species have another use for their lights, and that is to find and attract mates. Various species even have a "Morse code" of specific signals designed to attract only other individuals of their species. In some circumstances, a firefly species can be identified *only* by its flash pattern. Female fireflies of the genus *Photuris* go one step further by mimicking the signals of another species. This deceptive advertising is done to lure males of the other species, which the *Photuris* then eat. They then convert this firefly meal, which would be toxic to other animals, into a chemical defense.

The most dazzling insect light displays are produced by synchronous fireflies in Malaysia and the Smoky Mountains of North America. But whereas the Asian species flash in unison while perched in mangrove trees that line the shores of rivers and other aquatic habitats, the U.S. species flash in synchrony while flying.

Fireflies and glowworms produce their lights chemically in specialized cells in the abdomen. It is a complex reaction involving the enzyme luciferase acting on the chemical **luciferin** in the presence of the energy compound Adenosine Tri-Phosphate (ATP). Despite our ability to synthesize this reaction, like in the popular glowing sticks one sees in nightclubs, fireflies are still harvested in astronomical numbers for their light-producing chemicals. Hopefully, their lights will never cease blinking in nature, but conserving these insects should be a priority.

Good Vibrations

Single men and women in the dating scene are always hoping to receive "good vibes" from a potential mate, but many insects take that phrase literally. Few insects are capable of hearing the way we do, with auditory receptors, but they are much more advanced in detecting and identifying vibrations through objects and in the water. Insects can tell if a given tremor is a potential threat, an inconsequential breeze or wave, or a member of its own species. A wide array of insects from treehoppers to stoneflies to lacewings communicate by stomping out species-specific, rhythmic messages. They may do so by pounding with their feet or by banging their abdomen or another body part against a substrate. Soil, twigs, and other surfaces work just fine to transmit the intentions of the sender and reach the receiver without need for translation. Watching male jewel beetles like those in the genus *Chrysobothris* thump their rear ends vigorously on the surface of a log is quite entertaining. Sometimes it results in a multi-beetle pile-up atop a less-than-enthusiastic female.

Deathwatch beetles in the family **Ptinidae** are the "head-bangers" of the insect world. The tiny beetles live in tunnels in dead, solid wood. The male smacks his face against the floor of his tunnel to communicate his location to females. The knocking or ticking noise is audible enough to convince a person that their room or furniture may be haunted; and folklore has it that the noise of deathwatch beetles in a home portends an imminent death.

Communication by vibration can be used for purposes other than courtship and mating. Many insects generate distress calls or alarm signals when confronted with a potential threat. This is usually among members of their own kind, even their own family in the case of mother treehoppers (family **Membracidae**) herding their nymph offspring when danger approaches. Occasionally, the vibrations are meant for other species. Some caterpillars are known to "whistle" for ants to come to their defense when a parasite or predator gets too close. The caterpillar rewards its ant guards with a sweet liquid it exudes from glands in its body.

Drummer Boys/Percussion

Who does not like the drummer in a rock band? Percussion instruments are the heart and soul of many a musical ensemble, but there are drummers among the insects, too. The only insects that produce sound with a true percussion....instrument....are the cicadas, true bugs in the family **Cicadidae**, which are related to spittlebugs, leafhoppers, and planthoppers.

Males of most cicada species have flexible membranes inside the first segment of their abdomen, one on each side. These organs are called **tymbals**, and they are operated by muscles that contract and cause ribs on the tymbals to flex inward, making a sharp sound. Between contractions the tymbals' ribs snap back to their original position, again making a sound. This rapid (300-400 contractions per second) and repeated action produces sounds that reverberates inside a tracheal air sac that takes up most of the rest of the abdomen. Each chamber is covered by a "lid" called an operculum, one on each side of the underside of the abdomen. Noise levels produced by some cicada species can exceed 120 decibels at close proximity; a level similar to the sound of an ambulance siren. How a cicada gets so much bang for its tymbals' buck is still a bit of a mystery.

Australia's Bladder Cicada, *Cystosoma saundersii*, takes the prize for having the largest resonating chamber, which is contained in its enormously bloated abdomen. This species sings at night in the forests of Queensland, where its uniform green color hides it well among the lush vegetation.

Ironically, not all cicadas generate mating calls this way. There are many kinds of "mute" cicadas that are "wing-bangers". They simply strike the thickened leading edge of their front wing against their body, or whatever object they are perched on. This tapping sound is still surprisingly audible, but vastly quieter than the "harvest flies" and "dog day cicadas" that add an acoustic sizzle to hot summer and autumn days.

Fiddling Around

Anyone familiar with the fictional insect in *The Cricket in Times Square* knows that crickets (**Orthoptera: Gryllidae**) make their sweet music by drawing one wing over the other like a violinist plying her bow across the strings. Actually, that is not far from the truth. Generating sound by rubbing together two or more body parts is called **stridulation**, and although many insects are accomplished in this skill, none are better than katydids and crickets.

In the males of most of these species, the front pair of wings is highly modified to produce and broadcast calls that serve a variety of functions. Take field crickets and their kin, for example. Males can produce a loud "calling song" to attract a female, a "courtship song" to woo her once she is close, and a "rivalry song" to repel competing males. All this emanates from a "file" on the bottom edge of one wing being rubbed vigorously across a "scraper" on the opposite wing in the region that could be described as the "shoulder". Most crickets are "right-winged", which means that the file is located on the right front wing and overlaps the scraper on the left front wing. Katydids (**Orthoptera: Tettigoniidae**), by contrast, tend to be left-winged.

Creating the sound, however, is just the beginning. The front wings are also shaped and spread out to both direct sound and amplify it. Some tree crickets will even chew a hole in a leaf, and crawl partially through allowing them to use the leaf as a "megaphone" to amplify their call. Similarly, ground-dwelling crickets and subterranean mole crickets sing from the mouths of their burrows to help increase the impact of their songs. Despite their volume, cricket and katydid songs have a ventriloquist-like quality that make the singer extremely difficult to locate.

Grasshoppers (**Orthoptera: Acrididae**), while related to crickets and katydids, are not such virtuosos. Some species stridulate by rubbing the inside of the hind femur ("thigh" of the hind leg) against raised veins on their front wing. This usually results in short bursts of "song", not the sustained choruses and solos of crickets and katydids. Many band-winged grasshoppers, those with yellow, orange, red, or even blue hind wings that are only visible when the insect flies, can produce an aerial acoustic display at will. This behavior is called "**crepitation**", and exactly how it is accomplished remains poorly understood. The loud snap, crackle, and pop noises may be a result of rapid deflection of veins in the hind wings, or perhaps the front and hind wings rubbing against each other.

Take a night hike some evening, even if it's only in your backyard or a nearby park, to see how many kinds of crickets and katydids you can find by sound. There is more diversity than you might imagine, though most katydids are found in the deciduous forests of eastern North America.

References

Bennet-Clark, H. C. and A. W. Ewing. 1970. The love song of the fruit fly. Sci. Amer. 223: 84-92.

Bowers, W. S., L. R. Nault, R. E.Webb, and S. R. Dutky. 1972. Aphid alarm pheromone: isolation, identification, synthesis. Science 177: 1121-1122.

Buck, J. B. and E. M. Buck. 1976. Synchronous fireflies. Sci. Amer. 234: 74-85.

Carlson, A. D. and J. Copeland. 1978. Behavioral plasticity in the flash communication systems of fireflies. Am. Sci. 66: 340-346.

Carlson, A. D. and J. Copeland. 1985. Communications in insects. Quarterly. Rev. Biol. 60: 415-436.

Esch, H. 1967. The evolution of bee language. Sci. Amer. 216: 96-104.

Gould, J. L. and C. G. Gould. 1988. The honey bee. Sci. Amer. Lib., New York.

Kerkut, G. A. and L. I Gilbert [eds.]. 1985. Comprehensive insect physiology, biochemistry, and pharmacology, Vol. 1-13. Pergamon, Oxford, England.

Lloyd, J. E. 1965. Aggressive mimicry in *Photuris*: firefly femmes fatales. Science 149: 653-654.

Lloyd, J. E. 1971. Bioluminescent communication in insects. Ann. Rev. Ent. 16: 97-122.

Lloyd, J. E. 1979. Mating behavior and natural selection. *In*: Symposium: Sociobiology of sex. Florida Ent. 62: 17-23.

Snodgrass, R. E. 1923. Insect musicians, their music, and their instruments. Smithsonian Annual Report for 1923. 1923: 405-452

Walker, T. J. and D. Dew. 1972. Wing movements of calling katydids: fiddling finesse. Science 178: 174-176.

Wenner, A. M. 1964. Sound communication in honeybees. Sci. Amer. 210: 117-123.

Wigglesworth, V. B. 1972. The principles of insect physiology, 7th ed. Chapman and Hall, London.

Winston, M. L. 1987. The biology of the honey bee. Harvard Univ. Press, Cambridge, MA.

Architecture and Engineering

Tunnel Builders

One of the more ambitious tunnel-building efforts of all time was the "Chunnel", a tunnel for vehicular traffic beneath the English Channel. It serves to connect the island of Britain to France and the mainland of the European continent. This engineering feat is not without precedence, as many tunnels already carry incredible numbers of vehicles under rivers and estuaries in the United States and Japan. For example, the Holland Tunnel, which connects New York to New Jersey, was once the longest mechanically-ventilated underwater tunnel in the world, exchanging 3.7

million cubic feet of air per minute. It has served as a model for all other tunnels.

While perhaps less sophisticated, insects are also tunnel-builders, and they have been building their tunnels for a much longer period of time. **Embiidina**, or web-spinners, live inside tunnels that are formed from silk they spin from glands in the "foot" of each front leg. Ants, termites, and other soil-inhabiting insects usually construct tunnels by burrowing. In addition, termites tunnel through wood, and many subterranean termites also build extension tubes across inedible stone or cement to span the distance from their underground homes to above-ground wood or other source of cellulose. These enclosed pathways, which can be several feet long, protect the termites from water loss and predators like ants. Both ants and termites have salivary secretions that act like glue to bind and harden the organic fragments that make up their tunnels. Similar tunnel-enhancement is practiced by yellowjacket wasps that nest underground. Not only do they line their tunnels with mud, to which they add their saliva, but in autumn they also form mud "turrets" at the entrance to the nest burrow. Carpenter ants also line their earthen tunnels with sawdust. Unlike termites, carpenter ants do not eat wood; they merely excavate it to form chambers for their main and "satellite" nests, which may be a good distance away from the main nest.

Other tunnel-builders are found among the larvae of flies as well as caterpillars of butterflies and moths. Some members of these groups will tunnel, or "mine", through the central tissue (mesophyll) of leaves without damaging their upper or lower surface. The serpentine tunnels of these insects are so distinctive that if one knows the host plant and the cursive pattern of the miner, he or she can often determine the exact species of insect responsible.

Air Conditioning

On a hot summer day, an air-conditioned home is welcome relief from the heat and humidity. Elaborate machines cool, filter, and blow the air into the interior of a house or building. Some termites, meanwhile, accomplished essentially the same outcome hundreds of thousands of years before mankind. The most sophisticated system yet investigated is produced by the African termite *Macrotermes natalensis*. Nests of this species are constructed of soil particles molded into hollow mounds that can be 16 feet tall and 16 feet in diameter at the base. The walls are 16-23 inches thick and have numerous ridges on the outside that contain tunnels for circulation and diffusion of gases through the walls.

Termites are highly susceptible to drying out (dessication) and will die quickly if deprived of air saturated with moisture. The nests of *M. natalensis* usually have an internal humidity of 98-99%. The temperature at the center of the nest likewise varies no more than a few degrees. Remarkably, the carbon dioxide levels inside the nest are maintained at 2.7%, a low figure considering the nearly two million tiny occupants living and breathing within the mound. A nest with this great a population requires roughly 240 liters of oxygen daily.

The supply of oxygen and the regulation of temperature and humidity are both accomplished by the nest design. The center of the structure contains most of the workers and the fungus gardens that supply the colony with food. As the air in the center heats, it rises to a hollow at the top known as the "attic" and then travels back down the tubes in the ridges on the outside surface of the nest. As it moves through the tubes, carbon dioxide is replaced with fresh oxygen, and the air is cooled. Water is brought up from deep in the soil under the nest to maintain the high humidity. Food brought into the nest also contributes to the water supply.

Honey bees also use forced air evaporative cooling to regulate the internal temperature of the hive. Worker bees will align themselves at the hive opening, fanning their wings to ventilate the nest. Paper wasps have also been observed performing similar actions atop their exposed paper nest combs.

Food Storage

Food storage systems have been used by people for thousands of years. In many ways the development of effective food storage was an important element for the success and spread of human civilization. Today we have a diverse array of refrigeration, canning, and freezing methods to preserve food and extend its "expiration date".

The concept of food storage is not foreign to insects, either, especially social species like ants. The honey pot ants of the genus *Myrmecocystus*, common in the arid southwestern U.S., have specialized workers called "**repletes**" that serve as living storage containers for sweet liquids gathered by other workers. The abdomen of a replete swells to many times its normal size as it accommodates the input from its nestmates. Most of the liquid is waste products from other insects, especially aphids, scale insects, and related sap-sucking bugs that excrete copious amounts of "honeydew" in the course of filtering nutrients from their liquid diet.

When aphids and their kin are not plentiful, the honey pot ants turn to their repletes, found clinging to the ceiling of special nest chambers (wine cellars?) for sustenance. The repletes ration their supply of liquid food by regurgitating it into the mouths of worker ants that solicit them for it. This mouth-to-mouth food exchange is known as **trophyllaxis**, and it is an important means of communication and maintenance of social structure in ants, social bees, and wasps.

In addition to honey pot ants, all honey bees and most other social bees store honey in wax cells or other storage vessels. Honey is also stored by some social wasps (**Vespidae**), including honey wasps (*Brachygastra* spp.), paper wasps (*Polistes* spp.), and their relatives, inside the cells of their paper combs.

Liquids are not the only foods stored by insects. Harvester ants in the genus *Pogonomyrmex*, which make the pebbly mounds so conspicuous in the flat landscapes of the U.S. Great Plains and the arid Southwest, collect seeds to get their colonies through periods of food scarcity. Dung beetles create both "brood balls" of manure for their larval offspring and "food balls" to feed themselves as adult beetles. Solitary parasitoid wasps solve the problem of food spoilage in an entirely different way by storing one or more host insects in a burrow or mud nest. Solitary wasps only sting their victims into paralysis so that they do not die and rot before their larval offspring can begin feeding on them. Morbid, no question, but also ingenious.

Stonework

Masonry is a revered occupation and skill, but humans are not the only animals to have integrated stonework into their architecture. Many insects have, too. Many insects have, too. We will discuss caddisfly larvae in the chapter on construction materials later in the book, but there are others.

Leafcutter bee (**Megachilidae**) species in the genera *Anthidiellum* and *Dianthidium* construct free-standing, single-cell nests using plant resins and then reinforce and decorate them with small pebbles, sand grains, and other materials for camouflage and strength. In some cases, several such cells are grouped together on the surface of a stone, cliff face, or other natural object. Some species fasten their nests in the crotch of a bare twig, where it is more cryptic in appearance than you might suspect.

Certain thread-waisted wasps in the family **Sphecidae**, particularly the caterpillar-hunting species in the genus *Ammophila*, close their underground burrows with carefully-chosen pebbles that block the upper reaches of the burrow without allowing the plug to cave in on the vulnerable egg or larval wasp at the bottom of the near-vertical shaft. The pebbles will hold even as the mother wasp heaps more loose soil into the cavity to hide her nest. Some species will also use a stone to tamp down the earth to the same relative compactness as the rest of the ground around the burrow, making it nearly impossible for parasites like velvet ants to detect it.

Recycling

Some of our most successful recycling efforts have been during times of war when raw materials were at a premium. Scrap drives for paper and metal were common in the U.S. during World War II. Recycling has since become a vital necessity in curbing industrial and consumer waste in most major cities and towns. We are now more aware than ever of the toxins leaching into landfills and aquifers as a result of what we are throwing away.

Not surprisingly, insects have been "recycling" for millions of years. Among the more obvious insect recyclers are honey bees. Beeswax combs are used repeatedly in the yearly storage of honey, pollen, or brood (eggs, larvae, and pupae, the "baby" stages of bees). Even the wax used to construct combs is reused within a hive.

Many silk-spinning caterpillars at least partially consume their old silk and recycle it for future use. This is definitely the norm for spiders which, while not insects, are in the same phylum, Arthropoda.

Lots of insects, especially caterpillars, will eat their own shed exoskeletons immediately after molting. Not only does this recycle vital nutrients, but it also eliminates evidence the insect was ever there in the first place. Sharp-eyed predators like birds never fail to notice clues to the presence of insects, including those shed "skins".

The real recycling that insects do extends beyond themselves. Termites, cockroaches, crickets, dung beetles, carrion beetles, blow flies, and countless other insects recycle decaying organic matter into nutrient-rich soil that nurtures the growth of plants and starts the whole cycle of energy anew. The importance of these roles of decomposers cannot be overstated.

Sewing or Lashing

It was a real achievement when humans devised methods to attach materials securely together with pliant strands of plant fiber or sinew (animal tendons) to create shelters and clothing. Weaver ants (*Oecophylla* spp.) in Africa and tropical Asia employ primitive lashing and securing techniques, too. Most weaver ants are species that build nests by folding leaves into tent-like compartments secured by silk strands spun by their larvae. When several worker ants are able to pull two leaves together, other workers use the larvae like living sewing machines to stitch across the gap. Thousands of silken threads are needed to hold the leaves in place against the natural tendency of the foliage to spread apart. Most remarkable is the cooperation between the workers in pulling the leaves together. When leaves are too far apart, the ants form living bridges by grasping the "waist" of the nestmate in front of them with their jaws. These linked ants then allow other workers to crawl over them while tugging the leaf edges to narrow the gap.

The caterpillars of leaf-roller and leaf-tier moths use silk in a similar fashion. Caterpillars (and ant larvae, too, for that matter) spin silk from glands in their lower "lip". When you see a caterpillar waving its head rhythmically, it is likely laying down silk threads. When connecting leaf surfaces, the insect attaches one end of a thread to one surface and the other end of the thread to the opposing point. A taut result is achieved when the silk shrinks as it dries, pulling the leaf surfaces closer and holding them firmly in place.

Even the larvae of caddisflies (**Trichoptera**), the distant aquatic brethren of terrestrial caterpillars, employ silk in a variety of ways. Not the least of those uses is in lashing together plant debris, sand grains, pebbles, or other materials to fashion a "mobile home" designed to both disguise the insect as an inanimate object and also protect it from many underwater predators. Each family of caddisflies, or in many cases even each genus, has its own style of case-making architecture and design. Stare at the bottom of a shallow stream some time and you will probably start seeing "twigs" and linear or spiral collections of tiny stones moving across the streambed. No more can one consider "underwater basket weaving" a mythical class for college students.

Back on dry land, bagworms, the caterpillars of moths in the family **Psychidae**, festoon trees and shrubs with their cone-shaped, silken, and debris-decorated shelters. Their "bags" are thick and effective protections against almost all potential assailants. The pupa stage is also passed in this silken case that is firmly attached to a twig, fencepost, or other object. The adult female achieves sexual maturity without metamorphosis granting her wings. She never leaves her bag but emits a scent that attracts fluttering males to her shelter. It is her offspring, the tiny caterpillars that eventually emerge from her bag, that disperse the species. Each tiny caterpillar issues silk from its mouth until a breeze wafts it skyward and on to greener pastures.

Wallpaper

While people apply wallpaper mostly as a decorative element (your personal definition of "decorative" may vary) inside their homes, insects use the same concept for very different reasons.

Many solitary bees that nest underground need to keep the walls of their individual underground cells free from mold and as waterproof as possible. The female bee provisions a cell with a ball or loaf of "bee bread", pollen held together with nectar. She also lays an egg in the cell and then seals off the entrance from the main burrow tunnel. This marks the end of her parental care for that offspring. The larva that hatches from the egg will feast on the larder provided for it.

Female cellophane bees (**Colletidae**) in the genus *Colletes* secrete a natural polymer from a gland in their abdomens. Since the diet they provide for their larval offspring has a much more liquid consistency than the average ball of pollen, their natural wallpaper also serves as a kind of impermeable plastic bag to keep those nutrients contained, and fresh.

All such substances secreted by solitary bees and used in this fashion originate in the **Dufour's gland,** which produces chemicals that are deployed for a variety of tasks depending on the species, genus, or subfamily. All of the "advanced" members of the order Hymenoptera possess a Dufour's gland, and social species manufacture some communicative chemicals there.

Construction Materials

Few insects actually produce their own building materials, but honey bees are one exception. Worker bees possess hypodermal glands on the **sternites** (ventral abdominal segments) that produce beeswax used for construction of the combs in the hive. Workers consume about eight pounds of honey to produce one pound of beeswax. The combs are composed of adjacent six-sided cells that hold eggs and developing brood (larvae and pupae) and also serve as vats to store honey and pollen.

Insects also manufacture and secrete silk for construction. Among these silk factories are the Embiidina, or web-spinners, which have a silk gland inside the basal foot segment (basitarsus) of each front leg. The silk is used to form the tunnels in which they live, which are usually woven in existing cavities beneath bark on trees or in soil beneath stones or other objects.

Many moth caterpillars and other insect larvae produce silk, which is used to fashion cells (cocoons) in which they pupate. The silk of certain species of moths has been used to weave fabrics for human use. The best known of these is *Bombyx mori*, the Silk Moth (**Bombycidae**). This is the only truly domesticated insect species, one that is no longer able to complete its life cycle without human help.

The Carolina Leafroller (**Orthoptera: Gryllacrididae**), *Camptonotus carolinensis*, a relative of crickets that lives in much of eastern North America, is unique in having a silk gland in its mouth that it uses to spew silk threads to roll a leaf into a daytime shelter.

Even aquatic insects like the larvae of caddisflies and larvae of black flies (**Simuliidae**) produce silk. Caddisflies, depending on the species, make purse-like nets to filter food particles from swift water currents or weave together small objects to make a "mobile home". Black fly larvae each spin a cocoon in which to pupate.

The most unique and ingenious building material insects use is....are you ready for this? Their own feces. Leaf-eating beetles called casebearers (**Chrysomelidae**) use their fecal matter to craft a hard, protective capsule around their soft, vulnerable bodies. Caterpillars of certain tiny moths in the family **Batrachedridae** are also case-bearers that employ their **frass** (the scientific term for insect poop) for construction material. The lerp psyllids (**Hemiptera: Psyllidae**) of Australia construct a dome of crystallized honeydew, their liquid excrement, over themselves as protection. The Redgum Lerp Psyllid, *Glycaspis brimblecomei*, is one of two species that have become established in California and possibly Arizona, where it feeds on eucalyptus trees.

Pipe Liners

In the 1950s and 1960s, many paper mills used caustic chemicals for digestion, bleaching, and blending wood pulp. These chemicals were corroding the iron or steel pipes that conducted the flow of raw materials to the blending tanks and paper machines. One solution to the corrosion problem was the use of tongue-and-groove oak strips to line the pipes, preventing the mix from directly contacting the metal conduits.

Some insects have a similar problem, but of an anatomical origin. The food of some insects can cause physical abrasion of the lining of their alimentary tract. To prevent damage, their gut is protected by an intricate, chitinous lining called the **peritrophic membrane**. This laminate-like shield is produced by either epithelial cells at the front end of the gut near the cardiac valve or by the epithelium along the entire length of the gut wall. The membrane protects the sensitive gut wall from physical trauma and is also semi-permeable, such that only molecules of a certain size can pass through to be absorbed into the insect's body.

The rough, indigestible food particles continue to the hindgut, which is still enclosed by the peritrophic membrane, and are finally excreted as "packaged waste". Because the peritrophic membrane is eliminated as the "wrapping" around those fecal pellets, it must be continuously produced by the insect to maintain the healthy functioning of the GI tract.

The membrane serves another function in at least one beetle species. Larvae of *Ptinus tectus*, a spider beetle (**Ptinidae**), make their cocoons from the unbroken peritrophic membrane extruded from the larva's anus.

Reinforced Tubing

The radiator hose connected to most water-cooled engines is reinforced with a spiral or coil of steel to prevent the hose from collapsing under pressure as the coolant in the radiator expands and contracts. Similar reinforcement occurs in an insect's tracheal system.

An insect's respiratory system is made up of an extensive network of air tubes (**tracheae**) that branch until each of the hundreds of thousands of cells in the body is reached by a minute air tube called a tracheole. You read that correctly, each *cell* in an insect's body is serviced by its own individual breathing "hose".

The lining of the trachea is made of the same tough material (**chitin**) as the rest of the insect's exoskeleton. The main branches of the tracheae are even shed along with the exoskeleton when the insect molts. Those white stringy things on a shed "skin"? Those would be the old tracheae.

In addition to being built of sturdy proteins, the tracheae have spiral thickenings, called **taenidia**, in the lining (**intima**) of the tracheae. These give the tubes a ribbed appearance and prevent them from collapsing, much like the radiator hose in our introductory example. Engines and insects having something in common. Who would have thought?

Mud and Masonry

Whereas honey bees have mastered wax as an architectural medium, and social wasps have perfected paper as a construction material, many solitary wasps and other insects turn ordinary mud into masterpieces of engineering, and even art. Take the potter wasps (**Vespidae**) in the genus *Eumenes*, for example. They are related to yellowjackets, hornets, and paper wasps but are solitary, so each female fashions her own nest(s). She crafts a marble-sized urn, complete with a "neck" and "lip", and uses this receptacle to hold a cache of paralyzed caterpillars that she collects. She inserts a single egg inside, and once full of food for her future larval offspring, she seals the opening. She will repeat this process until she perishes. Look for her handiwork along the inside edges of recessed window frames.

More familiar is the work of mud dauber wasps of various species. The common Black and Yellow Mud Dauber (**Sphecidae**), *Sceliphron caementarium*, can make oval, individual cells of rather graceful shape or construct a conglomeration of cells that appears as if someone hurled a dirt clod and it stuck to the wall. They might be eyesores to us, but the thick cell walls are difficult for all but the most determined parasites and predators to break into. Inside each cell is a collection of paralyzed spiders feeding a single wasp larva. The Blue Mud Dauber (**Sphecidae**), *Chalybion californicum*, recycles the nests of its black and yellow cousin. It is also an enemy of black widow spiders.

The Pipe Organ Mud Dauber (**Crabronidae**), *Trypoxylon politum*, found in eastern North America, is a different wasp altogether. Each female makes a nest of parallel mud tubes that are sometimes more than six inches in length. Each tube is divided into a series of "apartments", each one housing a larval wasp that is feasting on paralyzed spiders. While the female is hunting, her mate may stand guard at the entrance of a tube to defend against parasites and predators as well as other males seeking a mating opportunity with the returning female.

Ladders

Ladders have been used by human cultures around the world since ancient times. They come in a seemingly endless assortment of shapes and sizes, from the ubiquitous six-foot stepladder to monsters rising hundreds of feet in the air on radio towers and smokestacks. Though humans have taken the ladder to new heights, insects did it first. Caterpillars of butterflies and moths move around on simple legs (prolegs), which are equipped with hooks or claws called crochets, which give caterpillars incredible traction when moving on a rough surface. Anyone that has allowed a large, living caterpillar to crawl on a finger or arm has felt the tiny hooks digging into her skin and may even have had difficulty removing the tenacious caterpillar from her arm. As powerful as the crochets are, they are useless on hard, smooth surfaces. When faced with such an obstacle, larvae of the gypsy moth (*Lymantria dispar*) deposit silk in rung-like fashion until it forms a simple ladder spanning the slippery area.

Spiders are not insects, but it is interesting to note that certain tropical orb-weaving spiders spin "ladder webs" of considerable length in order to trap moths. Like butterflies, moths have wings covered in scales that easily dislodge. A moth hitting a spider web may lose scales to the sticky silk threads but may still be able free itself. It's a sort of living Teflon organism to which spider silk does not stick. A ladder web is the spider's answer. A moth impacting a ladder web rolls down the length of the snare, having scales stripped from its wings as it tumbles and eventually leaving it with no scales left to shed. It then becomes trapped at the bottom of the web.

References

Akre, R. D., A. Greene, J. F. MacDonald, P. J. Landolt, and H. G. Davis. 1981. The yellowjackets of America North of Mexico. USDA Handbook No. 552.

Borror, D. J., C. A. Triplehorn, and N. F. Johnson. 1989. An introduction to the study of insects. Saunders, Philadelphia, PA.

Chapman, R. F. 1982. The insects: Structure and function, 3rd ed. Harvard University, Cambridge, MA.

Eiseman, C., and N. Charney. 2010. Tracks & Signs of Insects and Other Invertebrates. Stackpole Books, Mechanicsburg, PA.

Glancey, M., C. E. Stringer, Jr., C. H. Craig, P. M. Bishop, and B. B. Martin. 1973. Evidence of a replete caste in the fire ant, *Solenopsis invicta*. Ann. Ent. Soc. Amer. 66: 233-234.

Hölldobler, B. K. and E. O. Wilson. 1977. Weaver ants. Sci. Amer. 237: 146-154.

Hölldobler, B. and E. O. Wilson. 1983. The evolution of communal nest-weaving in ants. Amer. Sci. 71: 490-499.

Hölldobler, B. and E. O. Wilson. 1990. The ants. Belknap/Harvard Univ. Press, Cambridge, MA.

Kim, K. C. and R. W. Merritt [eds.]. 1987. Black flies: Ecology, population management, and annotated world list. Penn. State Univ., Univ. Park, PA.

Lofgren, C. S. and R. K. Vander Meer [eds.]. 1986. Fire ants and leaf-cutting ants. Westview, Boulder, CO.

Luscher, M. 1961. Air-conditioned termite nests. Sci. Amer. 205: 138-145.

Mackerras, I. M. [ed.]. 1970. The insects of Australia. Melbourne Univ. Press, Carlton, Victoria, Australia.

McKay, E. A. 1988. Tunneling to New York. Amer. Herit. Invent. Tech. 4: 22-31.

Richards, O. W. 1978. The social wasps of the Americas excluding the Vespinae. British Museum (Nat. Hist.), London.

Richards, O. W. and R. G. Davies. 1977. Imm's general textbook of entomology, 10th ed., Vol. 1. Wiley, New York.

Romoser, W. S. 1981. The science of entomology. Macmillan, New York.

Ross, H. H., C. A. Ross, and J. R. P. Ross. 1982. A textbook of entomology. Wiley, New York.

Strassman, J. E. 1979. Honey caches help female paper wasps (*Polistes annularis*) survive Texas winters. Science 204: 207-209.

von Frisch, K. 1967. The dance language and orientation of bees. Belknap/Harvard Univ. Press, Cambridge, MA.

von Frisch, K. 1983. Animal architecture. Van Nostrand Reinhold, New York.

Wheeler, W. M. 1910. Ants: Their structure, development and behavior. Columbia Univ. Press, New York.

Wilson, E. O. 1971. The insect societies. Belknap/Harvard Univ. Press, Cambridge, MA.

Chemistry

Antifreeze

Water, the necessity of life, has a downside. It expands in volume when it freezes as ice. Both cold-blooded organisms and machinery can suffer as a result of that. The solution? Antifreeze agents. Although antifreezes are used for a number of purposes, the best known is that associated with the water-cooled internal combustion engines of vehicles. During the winter the radiator and engine block must be protected by an antifreeze that lowers the coolant's freezing temperature. Without antifreeze the coolant will freeze and expand enough to damage the radiator and engine block. Most antifreezes contain ethylene glycol or similar substances, although wood alcohol (methanol) was used in the early days of the automobile.

Insects also have a freezing problem, which they have solved in essentially the same way. When winter arrives, many insects enter into a "cold hardy" stage. This can be the eggs, immature stages, or adults, depending on the species. Insects enduring a cold period must resist freezing since the formation of sharp ice crystals can pierce the walls of individual cells. This can prove fatal. The antifreeze insects produce in their bodies is glycerol, a type of alcohol. Glycerol lowers the freezing point of the blood (hemolymph) inside the insect and renders it cold hardy. In addition to producing antifreeze, the cold-hardy insects can actively excrete excess water to reduce the chance of a freezing injury.

Perfumes and Pheromones

A **pheromone** is a chemical secreted by one individual that causes a specific physiological or behavioral response in another individual of the same species. Usually, pheromones are classified into two groups depending upon the response elicited. Physiological pheromones are called **primers**, and they have a long-term effect. A well-known example is the queen pheromone of honey bees that, among other things, prevents the production of new queens in a normally functioning "queen right" colony. Other pheromones, called **releasers**, cause an immediate behavioral response in the receiving individual. An example of a releaser in the honey bee is the alarm pheromone that's associated with the bee's stinger. This volatile chemical, when deployed, both recruits more bees to the site of the first sting and instructs those warriors to sting the enemy, too.

Another common use of pheromones is as a sex attractant to lure individuals of the opposite gender for the purpose of mating. These "perfumes" are produced by females or males or, in a few species, by both sexes. Some male tiger moths (**Lepidoptera: Erebidae**) have enormous, tentacle-like glands they evert from inside their abdomens to better broadcast their scent. Males of other moths, and some butterflies, have "hair pencils" that also diffuse their pheromones more effectively. Meanwhile, male giant silkmoths in the family **Saturniidae** can home in on the faintest of female scents, flying a mile or more to find their dream date.

The study of pheromones essentially started with insects because insects are masters of the use of chemical signals. Even though insects have been the subjects of most pheromone research, it is now known that other animals, including humans, also produce pheromones. More "civilized" human societies have, however, largely replaced our natural odors with manufactured perfumes, colognes, musks, deodorants, and aftershave products.

Glue

Glues have an important role in our daily lives. Early glues, derived from plant and animal sources, were often frustratingly inadequate. Today, incredibly powerful and specialized adhesives are capable of adhering to almost any kind of material, porous or not. In addition to more mundane uses, glues are used to join human tissues during surgery and even to attach ceramic tiles to the (now-retired) space shuttle.

Not surprisingly, insects also manufacture and employ glues. The most common use of glues by insects is to affix their eggs to the substrate (physical surface) so they cannot be easily dislodged. Egg gluing takes many forms. The bright orange eggs of the Colorado Potato Beetle (**Coleoptera: Chrysomelidae**) are glued into place in small clusters on a potato plant, whereas the eggs of lacewings are individually suspended on long stalks glued to the surface of a leaf. Animal parasites such as lice (**Psocodea**) often glue their eggs on the pelage (hair, fur, feathers) of the host. The glue has to be incredibly durable to resist the scratching, clawing, and biting of the host in an effort to remove those eggs. The glues used by insects in attaching their eggs are produced by accessory glands, known as **collaterial** glands, which are associated with the insect's reproductive tract.

In the year 1990, the popular press reported that a construction engineer in Brazil had synthesized the salivary substances used by termites to glue their earthen nests together, and that this material was being used to build roads in the tropics. The substance was rumored to be as good or better than asphalt, and much cheaper.

Glue II

Ironically, another purpose for which we use glue is to trap insects. Fly paper, that tacky yellow ribbon you suspend from the ceiling to catch filth flies, is quite useful and effective, especially in restaurants where flies could potentially contaminate food. On the floor, glue boards entangle cockroaches and other pests, although snakes, lizards, and small mammals can become collateral damage. It is a horrible way for an animal to die, and homeowners should consider alternatives.

Naturally, predatory insects arrived at these uses of glue long before we did. The original flypaper may have been the strands of sticky, mucous-like silk deployed by the larvae of certain fungus gnats, family **Keroplatidae**, to ensnare small flies and other prey. The most famous example of these fungus gnats is the species *Arachnocampa luminosa*, the "glowworms" on the ceiling of Waitomo Caves in New Zealand. The larvae are bioluminescent, all the better to lure victims into the strands of glistening silk that they spin from glands in their mouths.

Certain assassin bugs have "sundew hairs" on their front and sometimes middle pairs of legs. These hairs help them hold onto struggling prey until it is subdued by a venomous bite. The assassin bugs (**Hemiptera: Reduviidae**) in the genus *Zelus* are best known for this. Each hair is associated with a gland in the leg that secretes a sticky substance to which almost anything will adhere. The insect grooms itself frequently to clean off debris so that its weapons are always at the ready.

Drug Use

A number of insects, and one mite, live on marijuana (*Cannabis sativa*), and other insects inhabit coca trees (from which cocaine is derived). Thus, insects made use of these narcotics long before humans discovered their pharmaceutical properties. In fact, flesh fly larvae (family **Sarcophagidae**) develop faster on the tissue of corpses that contain cocaine. This was somewhat unexpected but allowed forensic entomologist to more accurately predict the time of death of a victim during the investigation of criminal cases involving drug-related deaths.

Herbivorous (plant-feeding) insects of most types are able to either tolerate the chemical defenses of their vegetative hosts or actively use those chemicals to their own benefit. The classic example is the Monarch butterfly (**Lepidoptera: Nymphalidae**), the larvae of which feed on milkweeds, which contain potent toxins called **cardiac glycosides**. Monarch caterpillars are not only able to ingest those poisons, but they also sequester them into their own body tissues, thus rendering them inedible or unpalatable to vertebrate predators like birds. Monarch caterpillars advertise their toxic nature with a loud wardrobe of black, white, and yellow stripes. The chemical defense persists throughout metamorphosis, making the adult butterfly, clad in black and orange "warning colors", equally well-protected.

Most of the active compounds that make plants toxic to us are manufactured by the plants as natural insecticides to thwart those herbivores. The nicotine in tobacco is such an effective protection for the plant that few insects eat it, and the pesticide industry manufactures insecticides using the compound and its derivatives. Pyrethrum and related insecticides are made from dried chrysanthemum flowers. The list goes on and on.

Natural Polyesters (Bouncy Chemistry)

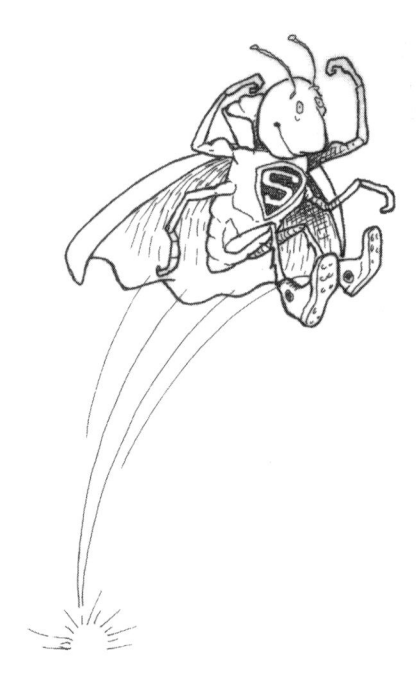

When human chemists found a material that could efficiently store the mechanical energy of compression, it was manufactured into small "rubber" balls for children. These "super balls", which bounced to heights out of all proportion to their size, are extremely resilient. Insects arrived at a similar effect through a different chemical compound. Have you ever wondered how tiny fleas can jump so high? Insect physiologists studying the leap of the fleas (order **Siphonaptera**) discovered that these insects have a super elastic protein called **resilin** in the thorax above the hind leg. Resilin is able to store 97% of the energy of compression. The muscles of the thorax and legs compress this protein so that it rebounds with great energy when released, catapulting the flea to great heights. Storage of kinetic energy in resilin is also important in the functioning of the "click mechanism" associated with insect flight muscles as the tension that is released into the power stroke of the wings. Thus, resilin is an important compound in the highly efficient use of energy by fast-flying insects.

Surprisingly, fleas are not the best leapers in the insect world. That record belongs to the froghopper *Philaenus*, in the spittlebug family **Aphrophoridae**, which can jump vertically to 115 times its body length, or about 30 inches! Oddly, resilin in the froghopper can store at most one to two percent of the energy needed for jumping. The remainder is stored in a specialized portion of cuticle called the **pleural arches**. It is stiff, but flexible enough to bow for jumping. The acceleration of the jump means the insect faces a G-force of 400 gravities. In contrast, a human astronaut leaving the launch pad for a space shot into orbit endures a maximum of about 5 G's inside his or her rocket. There is no word on whether a froghopper passes out or vomits during its jumps.

Preservation Without Freezing

Human civilization has managed to advance itself in part because of innovations in food preservation and storage. Freezing is our most common way to avoid the spoilage of food, especially meats and other proteins. Freeze drying, a method of dehydration, is another technique, as well as irradiation for the sterilization of food.

Some insects have developed ingenious ways of extending the viability of stored food, including the fungicides (page 62) and plastic "wallpaper" (page 47) discussed in previous chapters. Solitary wasps have a different solution up their...stingers.

A female solitary wasp stings her prey in its thoracic ganglion, a nerve center that is responsible for insect locomotion via legs and wings. The wasp's venom effectively paralyzes her victim, rendering it immobile, but still alive. She is then able to cache her prey in a burrow, mud cell, or other cavity and lay an egg on her victim, knowing it will be unable to escape. The wasp larva that hatches can then consume the larder without the risk of it rotting. The prey is thus more or less a zombie, but, unlike the zombies popular in movies and television, these zombies don't feed on the living but rather are themselves consumed.

Not all wasps completely paralyze their prey. The jewel wasps in the family **Ampulicidae** are cockroach hunters. The female wasp stings her victim in its neck, which leaves it mobile but unable to dictate its own movements. The wasp prunes her victim's antennae and then uses them as reins, leading the roach to a previously-selected or excavated cavity where the wasp lays a single egg on it before entombing it alive. Morbid, yes, but effective.

Dyes

The colors of insects are produced in two basic ways. Some are the result of microstructures that reflect and refract light. The metallic colors of some bees, wasps, flies, beetles, and butterflies are prime examples, as are the psychedelic eyes of horse and deer flies. Most earth tones are the manifestation of pigments in the chiton (cuticle) of the insect exoskeleton. As insects age, they darken with the build-up of lipids (fats) and may even become greasy or oily.

Insects can also accumulate or produce chemicals as a result of their diet, and these compounds can themselves be quite colorful. Such is the case with the chochineal insects in the true bug family **Dactylopiidae**.

These small, sedentary insects are the white "fluff" you observe on the pads of prickly pear cacti in the deserts of the New World. Their diet of cactus sap is turned, in part, into carminic acid, the "cochineal" that gives the insect its common name. As the name implies, the chemical compound is a vivid red color.

Carminic acid probably functions as a defensive compound, rendering the creature distasteful to predators. It might also serve as a warning color to advertise the insect's unpalatable nature, but normally the insect is hidden under a blanket of white, waxy secretions that protect it from desiccation (drying out) and sunburn that would likewise make the insect less appetizing to predators.

Human beings are less deterred by such things, if only because we are not likely to eat the insect, but rather use it for some other purpose. The Aztecs in pre-colonial Mexico utilized the crushed insects as a dye, and the Spanish conquistadors quickly exploited this technique by processing the dye as a dry powder and exporting it back to Europe. The most famous example of the use of cochineal was in dyeing the coats of the American Revolution's Redcoats.

Today, aniline and other synthetic dyes have largely replaced this Aztec dye, but cochineal is still used commercially. We owe the bright reds of cranberry juice beverages to cochineal, for example. The cosmetics industry employs the substance in lipsticks, rouges, and a host of other makeup products, and the pharmaceutical industry uses cochineal to color pills and ointments. It is also still used as a natural fabric dye, especially on fibers of animal origin.

Insect galls from plants also have a history of use as dyes. "Turkey red" is one example. Gall extracts have been used to dye wool and animal hides and also as inks.

Sunblock

As the ozone layer in Earth's atmosphere continues to erode and exposure to the sun's ultraviolet rays become ever more unavoidable, the application of sunscreen to one's skin is more important than ever. Although in the insect world there is no such thing as "tanning", many desert insects have built-in protection.

The chiton layer of the insect exoskeleton is usually packed with melanin, the dark pigment that helps protect us mammals from the sun's harmful rays. These pigments may also be complemented by other features like waxy excretions or reflective scales or hairs. Insects may not get sunburned, but they are at great risk of overheating and dehydration. Many bees, wasps, flies, ants, and beetles that live in deserts are covered in silver hairs that reflect sunlight, keeping them cool and, more importantly, minimizing water loss.

Aphids, cicadas, and their relatives are often coated in waxy secretions that make them waterproof (or nearly so) and also reflect sunlight. Oddly, some desert cicadas (**Hemiptera: Cicadidae**), like those in the genus *Diceroprocta*, allow for water loss through pores, thus "sweating" for an evaporative cooling effect. Many desert darkling beetles, family **Tenebrionidae**, have fused wing covers (elytra) to reduce water loss and an extra-thick exoskeleton to boot. Some species run so fast across the sand that they lose heat by convection.

Behavior can also help insects from overheating. Standing on tiptoe (tip tarsi?) elevates the insect's body from hot surfaces like desert sand or rocks. Grasshoppers may even alternate lifting their feet to avoid singeing their "toes". Dragonflies adopt a posture called **obelisking**, in which they point their abdomens straight up to minimize the amount of body surface that is exposed to the sun.

Temperature regulation in insects is a delicate balance, and while a butterfly might seek shelter beneath a leaf during a hot afternoon, it will be leaning into the sun at daybreak, basking in the morning light to warm its wing muscles for flight.

References

Anderson, S. and T. Weis-Fogh. 1964. Resilin: A rubberlike protein in arthropod cuticle, pp. 1-65. *In* J. Beament, J. Treherne, and V. Wigglesworth, [eds.], Advances in insect physiology. Vol. 2. Academic, New York.

Beroza, M. 1971. Insect sex attractants. Amer. Sci. 59: 320-325.

Birch, M. C. [ed.]. 1974. Pheromones. American Elsevier, New York.

Burrows, M. 2006. Jumping performance of froghopper insects. J. Exp. Bio. 209: 4607-4621.

Chapman, R. F. 1982. The insects: Structure and function, 3rd ed. Harvard University, Cambridge, MA.

Duplaix, N. 1988. Fleas: the lethal leapers. Nat. Geographic 173: 672-694.

Goff, M. L., A. I. Omori, and J. R. Goodbrod. 1989. Effect of cocaine in tissues on the developmental rate of *Boettcherisca peregrinae* (Diptera: Sarcophagidae). J. Med. Ent. 26: 91-93.

Gould, J. L. and C. G. Gould. 1988. The honey bee. Sci. Amer Lib., New York.

Hadley, N. 1993. Beetles make their own sunblock. Nat. History 102:44-45.

Hefetz, A, H. M. Fales, and S. W. T. Batra. 1979. Natural polyesters: Dufour's gland macrocyclic lactones form brood cell laminesters in *Colletes* bees. Science 204: 415-417.

Jacobson, M. 1971. Insect sex attractants. Amer. Sci. 59: 320-325.

Jacobson, M. 1972. Insect Sex Pheromones. Academic, New York.

Kerkut, G. A. and L. I Gilbert [eds.]. 1985. Comprehensive insect physiology, biochemistry, and pharmacology, Vol. 1-13. Pergamon, Oxford, England.

Metcalf, C. L., W. P. Flint, and R. L. Metcalf. 1962. Destructive and useful insects, 4th ed. McGraw-Hill, New York.

Naumann, I. D., [ed.]. 1991. The insects of Australia. Cornell Univ. Press, Ithaca, New York.

Romoser, W. S. 1981. The science of entomology. Macmillan, New York.

Ross, H. H., C. A. Ross, and J. R. P. Ross. 1982. A textbook of entomology. Wiley, New York.

Rothschild, Y., Schlein, K. Parker, C. Neville, and S. Sternberg. 1973. The flying leap of the flea. Sci. Amer. 229: 92-100.

Schildknecht, H. 1971. Evolutionary peaks in the defensive chemistry of insects. Endeavour 30: 136-141.

Sherman, H. 1989. Polyester repair-man is really a bee. Agric. Res. 37: 18.

Shorey, H. H. 1973. Behavioral responses to insect pheromones. Ann. Rev. Ent. 18: 349-380.

Shorey, H. H. 1976. Animal communication by pheromones. Academic, NY.

Steiner, A. L. 1983. Predatory behavior of digger wasps (Hymenoptera: Sphecidae) VI. Cutworm hunting and stinging by the ammophiline wasp *Podalonia luctuosa* (Smith). Melanderia 41: 1-16.

Toolson, E. C. 1987. Water prolifigacy as an adaptation to hot deserts: Water loss rates and evaporative cooling in the Sonoran Desert cicada, *Diceroprocta apache* (Homoptera: Cicadidae). Physiol. Zool. 60: 379-385.

Torchio, P. F. and D. J. Burdick. 1988. Comparative notes on the biology and development of *Epeolus compactus* Cresson, a cleptoparasite of *Colletes kincaidii* Cockerell (Hymenoptera, Colletidae). Ann. Ent. Soc. Amer. 81: 626-636.

Torchio, P. F., G. E. Trostle, and D. J. Burdick. 1988. The nesting biology of *Colletes kincaidii* Cockerell (Hymenoptera, Colletidae) and development of its immature forms. Ann. Ent. Soc. Amer. 81: 605-625.

von Frisch, K. 1967. The dance language and orientation of bees. Belknap/Harvard Univ. Press, Cambridge, MA.

Wilson, E. O. 1963. Pheromones. Sci. Amer. 208: 100-114.

Winston, M. L. 1987. The biology of the honey bee. Harvard Univ. Press, Cambridge, MA.

Tools

Thermometer

Temperature affects nearly everything we do as humans. Because our enterprises and activities are so dependent on the weather, we have television channels and websites devoted exclusively to meteorology. The weather conditions reported in the news usually contain maximum (high) and minimum (low) temperatures for the day as recorded by very sensitive and sophisticated thermometers.

Insects, being ectothermic or "cold-blooded", are governed even more in their behavior by changes in temperature. Some species can be used as living thermometers themselves by the mathematically inclined among us. The rate of chirping by male crickets in the family **Gryllidae** certainly slows down at cooler temperatures and speeds up on warmer nights. One enterprising naturalist, Amos Dolbear, devised a formula to convert the rate of chirps by a cricket into the air temperature in degrees Fahrenheit or Celsius. Dolbear's Law states that the temperature in degrees Fahrenheit is equal to the number of chirps in one minute minus forty, divided by four, plus fifty:

$$T_F = 50 + \frac{(N_{60} - 40)}{4}$$

The above equation can be simplified as $T_F = 40 + N_{15}$ for the number of chirps in fifteen seconds. This formula works best with the Snowy Tree Cricket, *Oecanthus fultoni*, though no one knows which cricket species Dolbear used. The rate of chirps for male field crickets in the genus *Gryllus* varies not only with temperature, but also with the age of the insect, mating success, and other factors that make *Gryllus* a less-reliable temperature gauge.

Other insects can also be used as living thermometers. Human body lice will abandon a host when its body temperature reaches 104° F. However, if you have a fever that high, you will likely have other symptoms; and hopefully be in a hospital under a physician's care. If not, the lice will also abandon your corpse after it cools below normal body temperature.

The entire life cycle of some insects is synchronized with temperature. For example, the spring emergence of adult Codling Moth (**Lepidoptera: Tortricidae**), *Cydia pomonella*, the "worm" in the apple as a caterpillar, can be predicted by the accumulation of days when the temperature is above a known threshold temperature (**degree-days**). Orchardists can use this information to time the application of pesticides to the period of peak adult Codling Moth activity. This greatly increases the effectiveness of sprays and reduces the amount of pesticides that are used.

Social insects must maintain the correct temperature (and humidity) inside their nests to allow them to properly function. Honey bees aim for a consistent 85°F but will increase the temperature to 95° F for optimal development of the brood (eggs, larvae, pupae).

Clocks

Humans have been measuring time ever since the beginning of....well, time. Today we have nuclear clocks that far surpass sundials, Stonehenge, and the hourglasses from ancient history. Quartz watches were considered the most precise of wearable timepieces until they were replaced by digital watches and, currently, smart phones and other multifunctional, high-tech electronic devices.

Left to our own biological devices, our human bodies operate on biorhythms known as circadian rhythms (**circadian** = about one 24-hour day), just like any other organism, including insects. That is, we follow these natural patterns when we are exposed to a normal day-and-night cycle of light and darkness. Deprived of those stimuli, such as deep inside a cave, a non-adapted organism will lose all sense of when to eat, sleep, and perform other functions vital to survival.

Insects make use of internal clocks for regulating the timing of many of their survival strategies, including mating and overwintering in a state of suspended animation called diapause. Honey bees have accurate clocks that enable them to detect differences in the position of the sun to within a few minutes. However, bees do have trouble orienting at the equator at noon, with the sun directly overhead, but only for a few minutes. Ants also have sophisticated navigation systems that use very accurate biological clocks.

Tool Use

Tool use involves the manipulation of an inanimate object, not internally manufactured, to improve the animal's efficiency in moving some other object. Another definition states that a tool is an object separate from an animal's own body that is used to extend the animal's capabilities. Tool use was once considered the sole domain of humans, or at least the great apes, but we have since discovered that even "lowly" insects can use tools to their benefit.

Tool use in social insects was first documented in 1964 when worker "pavement ants", genus *Tetramorium*, were observed dropping soil particles down the entrance to a solitary bee nest to induce the bee to come to the soil surface. Once the bee emerged, more worker ants attacked and killed it. Since then, similar behavior has been reported by numerous entomologists.

Another type of tool use by ants involves manipulation of soil or other absorbent organic debris to soak up honey for transport back to the nest or elsewhere within the nest.

Among solitary insects, tool use has also been recorded. Nymphs of the Costa Rican assassin bug *Salyavata variegata* decorate themselves with debris for camouflage so that they can approach the entrance to a termite nest undetected. The bug kills a worker termite with potent saliva from its beak-like mouthparts and feeds, then does a remarkable thing. It uses the dead termite to lure another termite within range and then kills that termite as well. The assassin bug repeats this cycle until it is satiated.

Still other assassin bugs, in the subfamily Apiomerinae, smear sticky plant resins on their front legs. Amazingly, the gummy substances are not used like glue to aid in the capture of victims, but to attract bees. The plant resins are highly aromatic and are especially effective at luring stingless bees to within striking distance. The "bee assassins" in the genera *Amulius* and *Ectinoderus* are especially adept at preying on *Trigona* stingless bees in this manner.

Some species of solitary wasps in the genus *Ammophila* have been seen using a pebble as a tool to pack sand over the entrance to their underground burrows. Though analogous to the use of mechanical compactors that construction crews use to smooth and compact soil after excavations, for the wasps it is a matter of life and death. Their nests need to be well concealed from potential predators and parasites. The female wasp catches a caterpillar, paralyzes it with her sting, and carts it back to her nest. She then lays an egg on it and refills the hole, using a pebble to tamp down the earthen closure. At the bottom of the burrow, a larva eventually emerges from the egg and devours the caterpillar over several days. Once grown, the larva enters the pupa stage, during which it transforms into an adult wasp that eventually bursts out of the pupa and digs its way to freedom.

SCUBA

Human exploration of the oceans, and even rivers and lakes, would not be possible without the development of devices to allow for underwater breathing. Since very early times, people have attempted to carry breathable air with them below the surface of the water. Some of the first attempts were rather crude, with animal skins being used to hold a limited supply that enabled individuals to go down a bit deeper or stay submerged a bit longer.

It was the French who first harnessed compressed air technology to free divers from an above-surface air supply. Yves Le Prieur developed the first open-circuit self-contained breathing apparatus (SCUBA) in 1925. During the German occupation of France in 1942, Jacques Yves-Cousteau and Émile Gagnan perfected the "Aqua-lung", which is the basis for most recreational diving done today.

While many aquatic insects breathe oxygen through gills, especially in their immature stages, most adult aquatic insects employ something akin to a SCUBA system for freedom to swim many minutes without needing to replenish their oxygen supply. Some make use of air bubble breathing, in which the insect at the water surface "grabs" a bubble of air to carry with it as it dives. The oxygen in the bubble is gradually depleted, eventually forcing the insect to return to the surface for a new bubble.

Other insects employ a plastron, a more elaborate form of respiration. A plastron is a thin layer of gas usually held in place against the insect's breathing holes, called spiracles. Specialized **hydrofuge** hairs (or other modifications of the cuticle layer of the exoskeleton) hold the gas, acting as a gill that allows oxygen to diffuse into the plastron as it is used. Insects that employ this type of respiration usually do not have to restore their oxygen level with visits to the surface, so they can stay submerged indefinitely. Both bubble and plastron respiration are used by aquatic true bugs and beetles.

Fishing Nets

The development of fishing nets allowed human fishermen to catch greater numbers of fish with less effort. When nets were incorporated into traps, fish could even be caught in the absence of fishermen, freeing them to invest more time in other activities and return to the traps at their leisure. Fishing nets are still used extensively today. They range from hand-held nets, to larger nets that can be cast from shore, to industrial drift nets that are miles long and that may fatally ensnare non-target animals like sea turtles or dolphins.

Insects also use fishing nets, though their quarry is other insects as well as microscopic organisms they filter from swift currents. The larvae of many species of caddisflies, order Trichoptera, live in silken tubes or case-like retreats attached to the bottom of rocks or on the streambed. From these houses, they issue nets of varying designs. Some are trumpet-shaped, whereas others are funnel-like or sheet-like. The mesh of these nets is often surprisingly uniform; under high magnification, they look much like nets of human manufacture.

In the islands of the South Pacific, some indigenous tribes employ spiders to spin their fishing nets for them. While spiders are not insects, they are related under the phylum Arthropoda. Orb-weaving spiders in the genus *Nephila* are very large, tropical spiders that spin wheel-like webs of extraordinarily strong silk that is capable of ensnaring small birds or bats.

One of these spiders, when allowed to craft its web inside a frame, can create a net that is strong enough for tribespeople to use as a hand-held fishing net.

In the Solomon Islands, spiders in the genus *Cyrtophora* spin webs that the islanders harvest for use as lures and traps to catch narrow-mouthed needlefish. The silk, after being wound into a reflective lure, is suspended from a kite so that it skips across the ocean surface, attracting the attention of the needlefish. After biting, the fish become ensnared.

Woodworking Tools

Human carpenters have a variety of instruments at their disposal for crafting fine furniture and other wood products. Insects, on the other hand, have only their own body parts. Despite what would seem to be their limitations, wood-boring insects work efficiently, if not methodically.

The opposable mandibles (jaws) of carpenter ants, carpenter bees, flathead borers and roundhead borers (the larvae of jewel beetles and longhorned beetles, respectively), horntail wood wasp larvae, termites, and many other insects are strong enough to chew through wood of varying densities and to resist the wear of doing so. Soft-bodied termites and beetle grubs still have thickly-sclerotized mouthparts, and when the insect molts during its youth, it gets a fresh set of jaws. Molting is therefore the way the mandibles are sharpened. A human carpenter, by contrast, would need to manually sharpen a saw or chisel or replace it altogether.

Termites, some wood-boring moth caterpillars, certain wasp larvae, and beetle grubs actively consume wood, relying on a suite of protozoans found only in their digestive tract to convert the dense, nutrient-poor cellulose into compounds usable to the insect for growth and metabolism. Carpenter ants and carpenter bees, in contrast, chew holes in wood to make a living space for their colony and larval offspring. The food of carpenter ants is protein in the form of other insects, seed coatings, and other sources, plus nectar from flowers and honeydew secreted as a waste product by aphids, scale insects, and other true bugs. Carpenter bees feed on pollen in their larval stage and on nectar as adult bees.

Peel back the bark on a dead tree or a log, and you are likely to observe intricate patterns etched by various insect larvae and even holes that lead into the bole heartwood. The tunnels may be packed with sawdust-like material, the "frass" excreted from the rear of the insect as it feeds. Larvae of bark beetles in particular leave species-specific galleries that can be remarkably beautiful, if not the ultimately lethal to a drought-stressed or fire-weakened tree. Your electric router would be challenged to create something so artistic.

References

Alcock, J. 1972. The evolution of the use of tools by feeding animals. Evolution 26: 464-473.
Barber, J. T., E. K. Ellgaard, L. B. Thien, and A. E. Stack. 1989. The use of tools for food transportation by the imported fire ant, *Solenopsis invicta*. Anim. Beh. 38: 550-552.
Borror, D.J., C.A. Triplehorn, and N. F. Johnson. 1989. An introduction to the study of insects. Saunders, Philadelphia, PA.
Brockman, H. J. 1985. Tool use in digger wasps (Hymenoptera: Sphecinae). Psyche 92: 309-329.
Deligne, J. 1999. Functional morphology and evolution of a carpenter's plane-like tool in the mandibles of termite workers (Insecta: Isoptera). Belg. J. Zool. 129: 201-208.
Dolbear, A. 1897. The cricket as a thermometer. Am. Naturalist 31: 970-971.
Fellers, J. H. and G. M. Fellers. 1976. Tool use in a social insect and its implications for competitive interactions. Science 192: 70-72.
Fowler, H. G. 1982. Tool use by *Aphaenogaster* ants: a reconsideration of its role in competitive interactions. Bull. New Jersey Acad. Sci. 27: 81-82.
Frost, S. W. 1959. Insect life and insect natural history, 2nd ed. Dover, New York.
Gillott, C. 1980. Entomology. Plenum, New York.
Hölldobler, B. and E. O. Wilson. 1990. The ants. Belknap/Harvard Univ. Press, Cambridge, MA.
Joklik, W. K. and H. P. Willett [eds.]. 1976. Zinsser microbiology, 16th ed. Appleton-Century-Crofts, New York.
Lin, N. 1964-1965. The use of sand grains by the pavement ant, *Tetramorium caespitum*, while attacking Halictine bees. Bull. Brooklyn Ent. Soc. 59-60: 30-34.

McDonald, P. 1984. Tool use by the ant, *Novomessor albisetosus* (Mayr). J. N.Y. Ent. Soc. 92: 156-161.

McMahan, E. A. 1983. Bugs angle for termites. Nat. Hist. 92: 40-46.

Moglich, M. H. J. and G. D. Alpert. 1979. Stone dropping by *Conomyrma bicolor* (Hymenoptera: Formicidae): a new technique of interference competition. Behav. Ecol. Sociobiol. 6: 105-113.

Romoser, W. S. 1981. The science of entomology. Macmillan, New York.

Saunders, D. S. 1976. The biological clock of insects. Sci. Amer. 234: 114-121.

Saunders, D. S. 1982. Insect clocks, 2nd ed. Pergamon, New York.

von Frisch, K. 1967. The dance language and orientation of bees. Belknap/Harvard Univ. Press, Cambridge, MA.

Family Life & Society

Social Behaviors/Societies

We tend to think of ourselves as the only truly social animals with rules, conventions, and lifestyles. However, many other animals are social, and they often establish elaborate dominance hierarchies that govern their behavior. This ensures that the individuals at the top of the social structure are always the first in line for food, water, and mating privileges. In somewhat modified fashion, according to our social customs, these same rules govern human behavior.

Truly social insects are classified as "eusocial", defined as having a reproductive division of labor, an overlap of generations, and cooperation in caring for the young. Eusociality has evolved at least thirteen times in insects, including all the ants and termites, many of the bees, and some wasps. *Microstigmus comes*, a small wasp in the family **Crabronidae**, represents one of the more recent discoveries of sociality in wasps; meanwhile, students of bee behavior are still finding examples of eusocial behavior in various species.

Primitively eusocial bees include some of the sweat bees in the family **Halictidae**. Most of these are considered solitary, with each female excavating a burrow in the soil or decaying wood. In some species, the daughters of a female will remain to care for siblings and/or dig branches from the main burrow where they rear their own offspring.

Many normally solitary insects, from wasps to lady beetles, occasionally exhibit social behaviors. Males of some solitary wasps will congregate in loose or compact clusters to spend the night or to endure inclement weather. Lady beetles are well known for migrating vertically from valleys to mountain ranges during the hot summer, when their aphid prey is less abundant. The beetles form large, dense blankets of thousands of individuals during these periods of inactivity.

Dominance Hierarchies

A number of animals exhibit dominance hierarchies in which the dominant individual takes priority in nearly all social interactions, including feeding and mating. Chickens have a "pecking order" in which the alpha (top) hen has the right to peck all her subordinates and not be pecked in return. Cows have well-established butting orders, and the dominant cow is always in the lead when the herd returns to the barn, with the other cows following according to rank. Dominance hierarchies also exist in troops of baboons, in monkeys, and in wolf packs.

In many animals, status is determined and maintained by intimidation or sometimes by ritualized combat. These "fights" can be physically damaging to the individuals involved but are rarely fatal. The subordinate eventually assumes a submissive posture, ending the altercation. A familiar submissive posture is a young puppy rolling onto its back to expose its throat to a dominant dog (or a scolding human owner).

Insects also establish and maintain dominance hierarchies. One of the first reported was for bumble bees, genus *Bombus*. A more dramatic example may be the paper wasp *Polistes gallicus*. A scientist studying the division of labor among females in a nest realized that the behavior he was observing was similar to the dominance hierarchies seen in other animals. He discovered that while several females may cooperate in founding a nest, one becomes the reproductive dominant (a **gyne**, functioning much like a queen bee) and lays most of the eggs. Her subordinates are forced to repress their own reproductive potential and instead act as workers, caring for the gyne's offspring.

Male field crickets in the genus *Gryllus* establish hierarchies associated with territoriality. This is also true of male dragonflies of many species, male bot flies and even male butterflies.

Myrmecophila manni, a tiny cricket found exclusively in ant nests, establishes linear hierarchies by fighting. Cockroaches also have dominant-subordinate relationships that are determined through butting and pushing matches. Even though the subordinate usually crawls away uninjured, it may later die from stress. This non-specific stress is also a very important factor in human health. Just ask any teenager who is a victim of bullying.

Gift-giving

Exchanging gifts or giving a gift in recognition of a special occasion is a common custom in human societies. It may be surprising to learn that many insects also partake in this practice, most often as part of courtship and mating. Male insects often enhance their chances of reproductive success by offering some kind of gift to their potential mates.

Female insects typically have a great aversion to being touched, probably because physical contact is usually associated with a predator looking to turn the insect into a meal. When a male approaches a female for mating, he must overcome this aversion. The female must also be receptive to mating, or the male suitor will be rejected.

The males of many species of katydids, and some crickets, produce a gelatinous mass that encases the spermatophore that they transfer during mating. This protein-rich wad may represent a considerable percentage of the male's body weight, so it is a major investment; but through this transfer, the male greatly enhances the likelihood that it will be his sperm, and not that of a competitor that fertilizes her eggs. Tree crickets in the subfamily Oecanthinae go a step further. The male has glands at the base of his wings that produce a secretion the female loves. She feeds on the liquid while the spermatophore pumps sperm into her genital tract.

Balloon flies in the family **Empididae** are predators of other insects, especially other flies. Males, especially those in the genera *Hilara* and *Empis*, will capture prey and present it to a female to distract her during mating. Some *Hilara* species even wrap their gift in a balloon of silk produced by glands in their front "feet". Some males practice deception by spinning an empty balloon that they then offer to their mate. In either case, the balloon itself is bright white and a larger object than normal prey, creating an **"overoptimal"** or **"supernormal"** stimulus that the female finds irresistible.

Certain hangingflies, scorpionflies in the family **Bittacidae**, also engage in the presentation of prey by the males to lure and occupy females during mating. Like the balloon flies, the males typically suspend themselves from a perch with their front pair of legs and manipulate the prey item with the middle or hind pair of legs. The female hangs beneath her mate while she indulges in her free nuptial meal.

An apparently unique vegetarian gift-giver is the African seed bug *Stilbocoris natalensis* in the family **Rhyparochromidae**. The male secures a fig seed for his prospective lover and injects enzymes to begin the pre-digestive process. Males without seeds are not successful in mating. Let that be a lesson to you, fellas.

Caste

In some human societies, especially in India, populations of people are divided into castes by social status and occupation or function. A member of one caste cannot aspire to rise in society nor marry outside his or her class. While human cultures can choose whether to abide by such parameters, social insects are physically limited to their role in the colony.

Truly social (eusocial) insect colonies are governed by at least one matriarch (the queen), with subordinate females in a "worker" caste. The colonies of some ant species may have multiple queens, but their collective role is solely to mate with males (drones) and reproduce. The workers are generally smaller in size than the queens but may be further subdivided based on their physical attributes. Soldiers may have large heads and mandibles used in fighting; minors are small workers better suited for chores within the nest. The presence of these "multiple forms" is called polymorphism, and it aids dramatically in the division of labor. Some soldier ants are equipped with such exaggerated jaws that they cannot even feed themselves and must rely on more proportionate workers to feed them.

All castes are rigidly controlled by chemicals called pheromones that are emitted by the queen(s) to maintain order in the colony. Thus, there are airborne "instructions" constantly diffusing throughout the nest, keeping every individual on task. Drones, which are only produced periodically, are sent from the colony into the countryside along with new, winged queens to mate with the reproductive members swarming from neighboring, unrelated colonies.

Cannibalism

Cannibalism, the consumption of members of your own species as food, is an extreme and rare behavior in human civilization today that is limited to tribes in the remotest of locations. Historically, such behavior was not ordinarily associated with obtaining nourishment, but rather had religious or social significance, such as an attempt to gain the physical or spiritual attributes of the vanquished. The only other circumstance leading to cannibalism is in predicaments so dire that eating another human (usually already deceased) is the only means of survival. Perhaps the most infamous case of this was the Donner party, which found itself trapped on a California mountain pass by an early snowfall in the 1800s.

Cannibalism in insects is, by contrast, fairly common. For example, the larvae of lady beetles (**Coleoptera: Coccinellidae**) will consume adjacent eggs containing their brothers and sisters. Horse fly larvae (**Diptera: Tabanidae**) will also voraciously attack and consume their siblings. Cannibalism allows these larvae to obtain energy until they can find other prey, and it reduces competition for that other prey. This becomes particularly important when prey is relatively scarce. Female green lacewings (**Neuroptera: Chrysopidae**) lay their eggs atop thin stalks, which reduces the possibility of cannibalism by the larvae before they can disperse.

Queens and workers of some social insects will produce inviable eggs that are laid solely as food to nourish the brood (larvae) and other adult members of the colony. These are called trophic eggs. The queen of a newly-established leafcutter ant colony may feed her offspring on trophic eggs until the fungus garden matures enough to feed them. Workers of stingless bee colonies also lay small trophic eggs that are consumed mostly by the growing larvae but are also eaten by the queen.

Yellowjackets, social wasps in the family **Vespidae**, are notorious for pulling out dead and diseased larvae from the paper comb, chewing them to a pulp, and feeding them to the other larvae in the nest. At the end of the season, when the colony is in decline, this behavior becomes even more frequent.

Drug Use

A recurring theme in human society is the use and abuse of narcotics such as heroin and other opiates, marijuana, tobacco, and related substances that provide momentarily relief and distraction from our daily lives, but often at a horrible cost to our health, interpersonal relationships, and productivity. Interestingly, many insects are likewise addicted, but for different reasons.

A number of insects, and one mite, live on *Cannabis sativa*, the marijuana plant, and other insects feed on coca trees, from which cocaine is obtained. The very chemicals that we find "stimulating" are in fact designed to repel insects and other herbivores. Some insects have evolved to overcome those compounds and exploit plants that otherwise have few competing herbivores.

Even more remarkable, many insects and related invertebrates manufacture their own defensive compounds, adding to the "chemical warfare" that defines much of predator-prey interactions and associated evolution. The scientific discipline of chemical ecology was created in the 1960s by the late Dr. Thomas Eisner of Cornell University and several of his contemporaries. Over the decades they discovered many compounds unique to insects and still continue to do so. Many of these "bug drugs" could prove highly useful in pharmacology.

Some insects steal these compounds from each other. The male fire-colored beetle, *Neopyrochroa flabellata*, presumably acquires the potent chemical cantharidin by licking blister beetles (**Meloidae**) or false blister beetles (**Oedemeridae**), both of which produce the chemical for their own defense. The fire-colored beetle stores the cantharidin in accessory glands linked to his reproductive system, but spares a little to shunt to a special groove across his face. Upon meeting a female, she tastes the contents of this facial pocket and assesses whether or not he has a store of cantharidin. Yes? Then she allows him to mate, during which he empties his hazardous store into what Dr. Eisner calls her "love canal". The female will then coat her eggs with the cantharidin, affording her ova ample protection from most predators.

Fireflies manufacture compounds called **lucibufagins**, akin to the toxins produced by toads. So coveted are these chemicals that females of at least some species in the genus *Photuris* will respond to the flash signals of males in the genus *Photinus*. Rather than attempting to cross breed, the female *Photuris* will eat the male *Photinus* to ingest his lucibufagins.

Begging

All human societies have beggars of one sort or another on street corners beseeching others for food or money. It is no different with social insects. Because social insects usually have stores of food, other insects constantly try to exploit their larder. Perhaps the closest analogies to beggars are found among the myrmecophiles or "ant lovers", insects that live within colonies of ants. One example concerns small crickets of the genus *Myrmecophila* that live with thatching ants (*Formica* sp.). These crickets are able to mimic the antennal touch greetings that worker ants use to solicit food from one another. Think of it as knowing the keypad code on the door of a secure food locker. The cricket will cautiously approach a worker ant, tap the ant with its antennae in a sequence that ordinarily causes the worker to feed a begging nestmate. Should the cricket perform the behavior incorrectly, it is very likely to be attacked and killed.

Other examples of inquilines or uninvited "guests" in ant and termite nests include certain silverfish, cockroaches, and beetles, who are all adapted by behavior or body structure to appease their hosts while begging for food. It is a remarkable way of life.

Robbers

Insects, just like people in some segments of human society, are not above stealing from others. Some individual females of certain solitary wasps will swipe the prey of others of their own kind when the rightful owner turns her back to open her nest burrow. Sometimes a wasp will even dig open a nest once the other wasp has resealed it and departed.

There is a family of flies, the **Asilidae**, that are commonly known as "robber flies" because some species in the group are adept at divesting other predatory insects of their hard-earned prey. Despite the name, most species of robber flies are excellent predators in their own right and do not steal prey from other insects. They must return to a perch to eat, however, and this can make even the thieves vulnerable to theft. It is not uncommon to find an ambush bug or another predatory insect "sharing" the kill of a robber fly.

Another entire family of flies, much smaller and less distinguished than the robber flies, are the "freeloader flies" (**Diptera: Milichiidae**). They look much like the *Drosophila* "fruit flies" in your kitchen, but they are outdoor insects that are seen almost exclusively on other insects killed by assassin bugs, crab spiders, and other terrestrial invertebrate ambush hunters. The freeloader flies are apparently too small for the predators to bother shooing away, so they are allowed to lap up the leftovers.

More remarkable still are certain blow flies in the genus *Bengalia* that pilfer from the most intimidating insect predators of all: army ants. The flies wait beside raiding columns of the ants and simply grab the prey from the worker ants returning to the nest. Since the ants are flightless, the flies accomplish their thievery with relative impunity, simply flying away to eat at their leisure. The flies exploit foraging carpenter ants in the same manner.

The kissing bugs in the assassin bug subfamily **Triatominae** are not averse to occasionally sinking their beak-like mouthparts into another member of their species to get to the blood meal the insect ingested from a warm-blooded host. There is apparently no resistance or harmful aftereffects on the "donor".

Communes

Semi-social living groups among humans are known as communes, and they became wildly popular in the United States during the protest movements of the 1960s. Relationships in the commune tended to be very elastic and non-permanent among the commune inhabitants.

Some insects have a similar type of communal organization. Communal nesting is a loose association of individuals with very little sharing of resources, if any. Several kinds of bees and wasps form nesting aggregations that are not without benefit to all concerned. Alkali bees, *Nomia melanderi*, (**Halictidae**) are one example, as are some members of the bee genus *Anthophora* (**Apidae**). The huge cicada killer wasps in the genus *Sphecius* may also take over a yard or small public space, causing undo alarm to the people around them.

The primary advantage to such communal nesting is that the likelihood of any one nest becoming victimized by a predator or parasite is greatly reduced. Your neighbor's burrow is as likely to be broken into as your own. Cicada killer nesting sites have the added benefit of being protected by the territorial males that drive off almost any intruder in the course of defending their own potential mates from rival males.

Alkali bees are so valuable as pollinators of alfalfa that farmers even plant "bee beds" in the appropriate soil to encourage the bees to nest around their fields. Occasionally, the bees will lose track of their own burrow entrance and accidentally unload their pollen and nectar provisions in their neighbor's nest. Such is life in a commune.

Child Protective Services

The family is the most basic unit of mammals and birds, but even many "lower" animals like reptiles, amphibians, and fish demonstrate parental care of their offspring. It was once believed that very few insects exhibited such traits, but scientists are now finding this is not the case.

The burying or sexton beetles in the genus *Nicrophorus* are exemplary parents. The female, often with the aid of a male, transports a small animal carcass to an appropriate spot, then excavates beneath the corpse, sinking it into a shallow grave. The body is then skinned and fashioned into a "meatball" with a shallow crater at the top. Into this depression the female lays her eggs. Once the larvae hatch, she feeds them bits of meat until they can feed themselves. All the while she licks the meatball to apply bacteria- and mold-resistant substances that protect the meat from spoilage.

Sand wasps in the genus *Bembix* (**Crabronidae**) are solitary, so each female digs a burrow that terminates in one or more cells. Like a mother bird, she practices progressive provisioning, bringing each larval offspring food on an as-needed basis. In her case, she kills flies that she brings back to feed her young.

Many true bugs, such as stink bugs, assassin bugs, and treehoppers, will guard their eggs against tiny wasps that would otherwise lay their own eggs on the bug's clutch, as well as from predators that would gobble up the eggs in one fell swoop. Protection is even extended to the nymphs that hatch from the eggs, at least until the babies molt one more time and disperse to go about feeding themselves.

Still other insects are model mothers. The female Tasmanian sawfly species *Perga lewisii* (**Hymenoptera: Pergidae**) stands guard over her mass of roughly 80 eggs, then broods over the caterpillar-like larvae that hatch out until those offspring are strong enough to disperse over the eucalyptus foliage that they will be feeding on.

Even earwigs, order **Dermaptera**, protect their eggs and young nymphs. The female European Earwig, *Forficula auricularia*, will turn her eggs like a mother bird and lick them, perhaps to apply antiseptic substances to protect them from mold in the earthen chamber that serves as her nest.

Baby Food

Mammalian mothers provide special food to their young in the form of milk produced by specialized glands, from which the baby (babies) suckles. There is surely no such analogous example from the insect world, right? Well, not so fast.

Remarkably, the females of some flies rear a single offspring in what amounts to a "womb", the larva reaching a mature state inside the mother's body by feeding on nutritive fluid produced by the female. The female then gives birth to a full-grown larva that pupates immediately. The most famous examples of this life cycle are the African tsetse flies in the genus *Glossina*. These are the flies that transmit the microbes that cause "sleeping sickness" in humans and nagana in cattle. Most louse flies in the family **Hippoboscidae** also follow this bizarre form of metamorphosis. Louse flies are blood-feeding parasites of birds and, to a lesser degree, mammals.

Less extreme examples of "baby food" include differences in the diets of larval and adult bees and wasps. Young animals of most species require protein to grow. Hence, larval wasps are fed a diet of other insects or spiders, depending on the species of wasp. Larval bees are provided with stores of pollen. Meanwhile, adult animals fuel their energetic activities with carbohydrates. So, adult bees and wasps drink nectar and other sweet substances.

However, not all baby food is created equal. Future queens in a honey bee colony are fed a special substance called "royal jelly", which is regurgitated from glands in the nurse bee's **hypopharynx** (throat). While every larva in the colony receives at least some royal jelly, those larvae destined for royalty receive much more. This difference in nutritional input affects the expression of certain genes, triggering, among other things, the fully-developed ovaries and greater longevity exhibited by queen bees in contrast to workers and males.

Determining the Gender of Your Offspring

One of the eternal debates among the culture of our contemporary times is whether we should determine the sex of our babies in advance of their birth. Do we even need to? Amazingly, many insects are able to do just that through a bewildering array of reproductive strategies.

Some species of insects, most notably some walkingsticks (**Phasmatodea**) and aphids (**Hemiptera: Aphididae**), do not always have a male gender in a given population. Females are able to produce viable offspring without fertilization of their eggs, a phenomenon known as parthenogenesis. Very few vertebrates are capable of this, namely a few species of lizards.

Social wasps have another unique situation known as **haplodiploidy**, which means that males have half the number of chromosomes as females. For example, once she mates, a queen yellowjacket can produce female or male offspring. In catastrophic instances when a colony loses its queen, worker yellowjackets, which are female, can lay eggs that produce males. The worker wasps are not capable of mating, so unlike the queen, who has two sets of chromosomes, they have only one set to invest in reproduction.

Aphids of many kinds, along with gall wasps (**Hymenoptera: Cynipidae**) and a few other insects, practice an "alternation of generations" whereby one generation is the product of **parthenogenesis** (asexual reproduction) and the second is a product of traditional sexual fertilization. This allows for exploitation of different host plants, or parts of host plants, and other advantages not enjoyed by insects that rely on only one form of reproduction.

References

Akre, R. D. 1982. Social wasps, pp. 1-105. *In* H. R. Herman [ed.], Social insects, Vol. 4. Academic, New York.

Borror, D. J., C. A. Triplehorn, and N. F. Johnson. 1989. An introduction to the study of insects. Saunders, Philadelphia, PA.

Borror, D. J. and R. E. White. 1970. A field guide to the insects of America North of Mexico. Houghton Mifflin, Boston, MA.

Catts, E. P. 1967. Biology of a California rodent bot fly *Cuterebra latifrons*. Coq. J. Med. Entomol. 4: 87-101.

Dixon, A. F. G. 1959. An experimental study of the searching behavior of the predatory coccinellid beetle *Adalia decaempunctata* (L.). J. Anim. Ecol. 28: 259-281.

Englemann, F. 1 970. The physiology of insect reproduction. Pergamon: New York.

Ewing, L. S. 1967. Fighting and death from stress in a cockroach. Science 155: 1035-1036.

Gould, J. L. 1982. Ethology: The mechanisms and evolution of behavior. Norton: New York.

Henderson, G. and R. D. Akre. 1986. Biology of the myrmecophilous cricket, *Myrmecophila manni* (Orthoptera: Gryllidae). J. Kansas Ent. Soc. 59: 454-467.

Henderson, G. and R. D. Akre. 1986. Dominance hierarchies in *Myrmecophila manni* (Orthoptera: Gryllidae). Pan-Pac. Ent. 62: 24-28.

Hill, K. and A. M. Hurtado. 1989. Hunter-gatherers of the New World. Amer. Sci. 77: 436-443.

Hölldobler, B. and E. O. Wilson. 1990. The ants. Belknap; Harvard Univ. Press, Cambridge, MA.

Huber, P. 1802. Observations on several species of the genus *Apis*, known by the name humble-bees, and called Bobinatrices by Linnaeus. Trans. Linn. Soc. Lond. Zool. 6: 214-298.

Ito, Y. 1960. Territorialism and residentiality in a dragonfly, *Orthetrum albistylum speciosum* Uhler (Odonata: Anisoptera). Ann. Ent. Soc. Amer. 53: 851-853.

Jacobs, M. E. 1955. Studies on territorialism and sexual selection in dragonflies. Ecology 36: 566-586.

Kato, M. and L. Hayasaka. 1958. Notes on the dominance order in experimental populations of crickets. Ecol. Rev. 14: 311-315.

Kessel. E. L. 1955. The mating activities of balloon flies. Syst. Zool. 4: 997-1004.

Kistner, D. H. 1982. The social insects' bestiary, pp.1-244. *In* H. R. Herman [ed.], Social insects, Vol. 3. Academic, New York.

Matthews, R. W. 1968. Nesting biology of the social wasp *Microstigus comes* (Hymenoptera: Sphecidae: Pemphredoninae). Psyche 75: 23-45.

Michener, C. D. 1970. Apoidea, pp. 943-951. *In* E. F. Riek, Hymenoptera, pp. 867-959. *In* I. M. Mackerras [ed.], The insects of Australia. Melbourne Univ. Press, Carlson, Victoria, Australia.

Miller, N. C. E. 1971. The biology of the Heteroptera, 2nd ed. Classey, Hampton Middlesex, England.

Pardi, L. 1948. Dominance order in *Polistes* wasps. Physiol. Zool. 21: 1-13.

Richards, O. W. and R. G. Davies. 1977. Imm's general textbook of entomology, 10th ed., Vol. 1. Wiley, New York.

Seyle, H. 1973. The evolution of the stress concept. Amer. Sci. 61: 692-706.

Thornhill, R. and J. Alcock. 1983. The evolution of insect mating systems, Harvard, Cambridge, MA.

Wilson, E.O. 1971. The insect societies. Belknap/Harvard Univ. Press, Cambridge, MA.

Arts and Entertainment

Acrobats

Traditional and contemporary circus shows feature various acts in which human acrobats demonstrate their remarkable skills at tumbling or balance. Insects may have been the original acrobats, though, and there are some striking examples.

Flies of the family **Piophilidae** are commonly known as "cheese skippers". The larvae are occasional pests of dried meats and cheeses, as they feed on the fats in those foods. The "skipper" part of the name comes from their ability to catapult themselves through the air, sometimes up to several inches. They accomplish this by bending their body nearly double, grasping their rear end in their mandibles, tensing their muscles, and then letting go. As their body straightens and strikes the substrate (surface of whatever they are on), they are thrown high into the air. This probably startles would-be predators and facilitates escape.

Mexican jumping beans, once sold as dime-store novelties, are seeds of *Sabestiania* plants infested by the caterpillar of a moth, *Cydia deshaisiana* (**Tortricidae**). The larva, enclosed in the bean, will thrash about when the seed is warmed in a human hand or placed in sunlight.

A phenomenon similar to the jumping bean, the jumping oak galls, is caused by tiny gall wasps in the genus *Neuroterus* (**Cynipidae**), on the underside of North American White Oak leaves. The galls are deciduous, so they fall off the leaves and onto the ground in mid-summer. The tiny spheres bounce on occasion when the wasp larva inside moves violently.

There are tightrope walkers in the insect world, too. Certain wasps that are predators or parasites of spiders have learned how to walk on spider webs to reach their victims without becoming entangled. Other wasps that are parasites of leafroller caterpillars will either reel in the caterpillar's escape thread (see "Bungee Jumpers") or walk along the line to get to the larva.

Finally, there are the masters of the flying trapeze. Well, sort of. Scientists have discovered the intentional gliding behavior of "turtle ants" in the genus *Cephalotes*, which live high in the canopy of tropical rainforests. They take a flying leap and then control their descent so that they land back on the tree trunk instead of the ground. This is helpful considering such forests are flooded for a good portion of the year. Other arboreal (tree-dwelling) ants behave similarly.

Bungee Jumpers

Some human adrenaline junkies claim that bungee jumping is one of the most exhilarating experiences of their lives. Tie an elastic cord around your ankles and jump from a tall bridge. The cord will safely stop your descent. What a rush! Imagine if your life depended on this skill and you'll have some idea how some insects survive.

Caterpillars of the Apple-and-Thorn Skeletonizer, *Choreutis pariana*, (**Lepidoptera: Choreutidae**) frequently dive out of trees when they are threatened by predators, only to be saved by their "bungee cord" of strong silk spun from salivary glands in their mouths. Many of the leafroller caterpillars in the moth family **Tortricidae** have similar escape strategies: rocketing out of one end of their leafy retreat if discovered by a bird, wasp, or other enemy. A caterpillar you see suspended seemingly in mid-air along a woodland trail is likely one of these larvae that has narrowly escaped death.

Not all bungee jumps end successfully, however. Birds sometimes spot the suspended caterpillar and swoop down to snatch it up. Some persistent female parasitic wasps will either reel in the silken cord or slide down it to reach the caterpillar. The wasp then lays an egg inside the larva so that her offspring can feed as an internal parasite on the caterpillar.

Larvae of black flies (family Simuliidae) live in the strong currents of fast-moving streams. They employ a silken safety cord or "bungee" to keep from being swept away when they change positions. Normally, the larvae are anchored to an underwater rock by tiny hooks on their posterior that are lodged in a silken pad they spin.

Humming

We sing songs and hum through our mouths, but few insects do. The Death's Head Hawkmoths of the genus *Acherontia* (Sphingidae) are an exception. One of these moths was featured in the motion picture *Silence of the Lambs* and is even depicted on one of the promotional posters. The adult insect has a short proboscis, through which it draws in air. The in-rushing current causes the flap-like **epipharnyx** in the moth's "throat" to vibrate and produce a pulsed airstream and accompanying low-pitched hum. When the air is expelled, it creates a high-pitched whistle.

The moths make these sounds when handled, and the noises may act as a warning to potential predators. The same apparatus and behavior are reputed to produce sounds that mimic the piping sounds of a virgin honey bee queen. Since the moths are well-known for raiding hives to get to the honey stores, this is not a far-fetched hypothesis. Still, many moths fail in their thieving attempts and are killed by the worker honey bees.

Dancing

One of the most popular avenues for human entertainment is dancing. While we pursue the activity for sheer pleasure, or professional competition, insects do it for specific purposes.

The most famous example of insect dance is the "dance language" employed by honey bees. Foraging workers that find an abundance of nectar- and pollen-rich flowers will return to the hive and communicate the location of the flower patch to other workers. They do this through a choreographed figure-eight movement known as the "waggle dance". It is the orientation of the dancing bee on the comb that gives the direction of the resource in relation to the sun, and it is the intensity of the "wiggle" as the bee crosses the middle of the "8" that indicates the approximate distance from the hive. The waggle dance is also used to broadcast the location of water sources and potential new nest locations during swarming events. A less complex "round dance" is used for directions to resources closer to the hive.

Pomace flies in the family **Drosophilidae**, the tiny gnat-like flies in our kitchen that we call "fruit flies", engage in courtship dances. The male flicks his wings in an exaggerated fashion as he circles a potential mate. There is often a "singing" compliment as he vibrates his wings. Some true fruit flies (**Tephritidae**) dance as well, as do many other kinds of related flies. Many species have the wings decorated with dark patterns that add to the pageantry. Look for the little Peacock Fly, *Callopistromyia annulipes*, (**Ulidiidae**) performing on a fence rail near you. The male holds his wings together, perfectly vertical, as he sidles to and fro to impress the ladies.

Some species of robber flies (**Asilidae**) have courtships involving posturing and gesturing by the males, often performed while hovering. This is especially true for *Cyrtopogon* species, the males of which are adorned with colorful hairs on the abdomen, legs, and/or feet, which they display prominently.

Male mayflies, order **Ephemeroptera**, perform aerial ballets while gathering in airborne **leks** to attract the attention of females. The males rise in the air, then descend slowly, watching for females to enter the swarm from above. Many male true flies in the suborder **Nematocera** (gnats, midges, mosquitoes, and kin) also gather in dancing flight swarms, often over some prominent object. That is why the tallest member of your lakeshore hiking party is likely to be followed by a swarm of male midges (**Chironomidae**).

Sculpture

Art museums and public spaces are often adorned with sculptures and statues that inspire us, provoke us, or make us smile with their whimsy. Some insect architecture could easily be interpreted as sculpture that is artistic in nature but functional, too.

Termite mounds are prominent features of the landscape in the tropics and in the savannahs and deserts of Africa and Australia. Their scale is breathtaking, especially considering the small size of the insects that create them. The mounds vary in form from twenty-foot towers built to air-condition the occupants to the blade-like "compass mounds" of the Australian *Amitermes meridionalis* (**Blattodea: Termitidae**). These mounds are always oriented with the narrow axis running north and south to allow the broadest surfaces to capture the sun's rays from the east and west. The mushroom-shaped mounds of African *Cubitermes* termites help shed rain during the wet season. *Nasutitermes* species that live in mangrove swamps build aerial castles fabricated from their own fecal material. Some species of ants make similar "carton" nests in trees, or as shelters for the aphids they protect in exchange for honeydew, the aphids' sweet liquid waste product.

Some solitary wasps create wonderful sculptures, too. Potter wasps in the subfamily **Eumeninae** (**Vespidae**) fashion perfect mud urns, complete with a "neck" and "flute" into which they deposit paralyzed caterpillars as food for their larva. Each vase or pot will rear a single wasp larva. When she's finished filling it with food and laying an egg, the female wasp seals the vessel and goes about constructing another one. The Pipe Organ Mud Dauber, *Trypoxylon politum*, is a common eastern North American wasp. Each female constructs one or more linear mud tubes that she subdivides into cells. Each cell is then filled with paralyzed spiders as food for a single larval offspring. When two or more of these tubes are built adjacent to each other, it gives the impression of an old-fashioned pipe organ.

The paper nests of tropical social wasps can take the shape of a giant, ridged potato chip (*Synoeca* spp.), frosted hexagon or star (*Chartergellus* spp.), bell or suspended sphere (*Polybia* spp. and kin), or any number of other fanciful objects.

References

Atkins, M. D. 1980. Introduction to insect behavior. MacMillan, New York.

Borror, D. J., C. A. Triplehorn, and N. F. Johnson. 1989. An introduction to the study of insects. Saunders, Philadelphia, PA.

Borror, D. J. and R. E. White. 1970. A field guide to the insects of America North of Mexico. Houghton Mifflin, Boston, MA.

Haskell, P. T. 1964. Sound production, pp. 363-608. *In* M. Rockstein [ed.], The physiology of the insecta, 2nd ed., Vol. 11. Academic, New York.

Metcalf, C. L., W. P. Flint, and R. L. Metcalf. 1962. Destructive and useful insects, 4th ed. McGraw-Hill, New York.

Oldroyd, H. 1964. The natural history of flies. Norton, New York.

von Frisch, K. 1966. The dancing bees: An account of the life and senses of the honey bee. Methuen: London.

von Frisch, K. 1967. The dance language and orientation of bees. Belknap/Harvard Univ. Press, Cambridge, MA.

von Frisch, K. 1983. Animal architecture. Van Nostrand Reinhold, New York.

Wynne-Edwards, V. C. 1962. Animal dispersion in relation to social behavior. Oliver and Boyd, London.

Department of Defense & Warfare

Chemical Weapons

Napalm, mustard gas, phosgene, agent orange, and sarin are among the worst chemical compounds ever deployed by humanity in wartime. So heinous are these weapons that they are almost unanimously banned under any circumstances by international agreement. If only insects were so respectful of their adversaries.

Insects are masters at using noxious or stinging chemicals for their individual and collective defense. One example is formic acid, which some ant species spray from glands in their abdomens. Indeed, **Formicidae**, the family name for ants, is derived from this acidic substance. If it is disturbed, a colony of thatching ants (*Formica* spp.) will produce a gaseous or misty liquid pall of formic acid above the exterior of the nest. This pungent compound, which "takes the breath away" from most mammals, is equally effective against many invertebrates such as other ants, beetles, and millipedes.

Some birds take advantage of this defense response and engage in "anting", a behavior in which the bird allows itself to become covered by enraged ants that spray their acid and bite, thereby killing (or at least incapacitating) lice, mites, and other ectoparasites on the bird. Historically, even people have placed their furs and blankets on ant mounds for the workers to remove lice and fleas.

A truly stunning and spectacular form of chemical defense is practiced by various ground beetles (**Coleoptera: Carabidae**) collectively known as "bombardier beetles". The best-studied of these belong to the genus *Brachinus*, which produce a hot, (212° F) spray of chemicals called **quinones** when they are disturbed or attacked. Quinones are so reactive that they exist in the beetles as inert precursors in separate glands, mixed in an anal "reaction chamber" from which they exit with explosive force in rapid pulses. An audible "pop" accompanies the discharge, and anyone handling one of these bombardiers can attest to the scalding heat generated.

Many other beetles employ chemical defenses. Benzoquinones are common to the defenses of a variety of beetles, although adlehydes, ketones, and even steroids occur in some species. Steroids, which are only produced by diving beetles (**Dytiscidae**), are among the most complex of all defensive chemicals. When fish swallow dystiscids, the steroids cause the fish to regurgitate the beetle unharmed. Fireflies (**Lampyridae**) are full of **lucibufagins**, potent compounds akin to the toxins in toads' skin glands.

Another beetle to beware of is the blister beetle. All members of the families **Meloidae** and **Oedomeridae** contain an irritating chemical called **cantharidin**. Molest one of these soft-bodied beetles and it will bleed the chemical from body joints. The chemical can cause painful, scarring blisters on sensitive human skin and can be lethal when ingested. Livestock can be killed by eating blister beetles inside hay bales. The topical product

"Spanish fly", which many people believe acts as an aphrodisiac, is derived from the Mexican blister beetle *Lytta vesicatoria* and should be avoided at all costs.

Perhaps one of the most unusual uses of a chemical defense is the use of a spray by some female ground beetles to thwart the unwanted attentions of amorous males. The spray physically incapacitates the male so that the female can escape. Thus, we also have the first use of "mace" by a mere insect.

Camouflage

The need to remain unseen and undetected is of critical importance to humans who are hunters and to military personnel in times of war. Camouflage netting is often used by waterfowl hunters to conceal their blinds and by the military to disguise tactical installations. Virtual invisibility tools such as the B-2 bombers and more recent aircraft that utilize "stealth" technology to make the planes undetectable to radar are becoming increasingly complex.

Humans learned many of the principles of camouflage and cryptic coloration by observing animals, and especially insects, that use camouflage to their advantage. Intricate color patterns that match the background against which an insect is positioned also serves to break up the creature's outline. The effect may be enhanced by "decorative" protrusions of the insect exoskeleton that serve no other purpose. Lastly, insects have perfected the art of stillness and/or rhythmic movements that render them invisible to would-be predators.

One classic example of cryptic coloration concerns the Peppered Moth, *Biston betularia*, a member of the family **Geometridae** that is native to England but also found in other parts of Europe and has been introduced to North America.

Most individuals in a population of this species have a pale, dark-speckled color pattern, similar to the natural coloration of the bark, but a dark solid-colored form also exists. During the Industrial Revolution soot generated by the many coal-burning factories quickly accumulated on tree trunks where the moths rested by day, so light-colored moths quickly became visible to birds and other predators. By the end of the 19th century, in the United Kingdom, the dark form of the Peppered Moth was more common than the "normal", light-colored form. This case was termed "industrial melanism" for the natural response of the moth population to human impact. Since the early 1960s, with the establishment of regulations that have reduced air pollution markedly, the light form of the Peppered Moth has returned to its normal abundance.

Warning Colors

There is a uniform strategy for how to get the attention of human beings: Use red, neon yellow, fluorescent orange, or similarly impossible-to-ignore colors to communicate danger or hazard. Hunting jackets and vests worn by road construction crews come to mind. Red in particular seems to have a psychological effect on people by arousing caution or anger.

Insects also use bright colors to their advantage, for similar purposes. Many brilliantly-colored insects intentionally "advertise" their presence, in stark contrast to the usual camouflage. These flamboyant insects are often distasteful to predators and gain an advantage by making a spectacle of themselves instead of hiding. Some also have other defenses such as chemical sprays, venomous stings, or bites, so if an inexperienced predator still attacks, it will be met with severe consequences and quickly learn to associate the color and pattern with something unpleasant.

The insects' bold color combinations of red and black, black and yellow, orange and black, black and white, or even the solid metallic blue of the morpho butterfly are therefore aimed mostly at birds, mammals, and other vertebrates with keen eyesight and the memory to recall the connections between insect colors and defenses over time.

Many insects magnify the concept of warning coloration by gathering together. True bugs in the order **Hemiptera** frequently condense into loose groups, the better to advertise their odorous chemical defenses or distasteful nature due to the toxic plants they feed on. Immature stink bugs, shield bugs, leaf-footed bugs, and others exemplify this behavior; juvenile and adult cotton stainers, soapberry bugs, and giant mesquite bugs will often stay close together and even molt from one life stage to another in synchrony.

Not every example of warning coloration is backed up by a powerful defense. A good many completely vulnerable insects wear flashy patterns, too. Why? You'll have to read the chapter on mimicry to learn the answer. You can't say we don't leave you in suspense.

Radar/Sonar

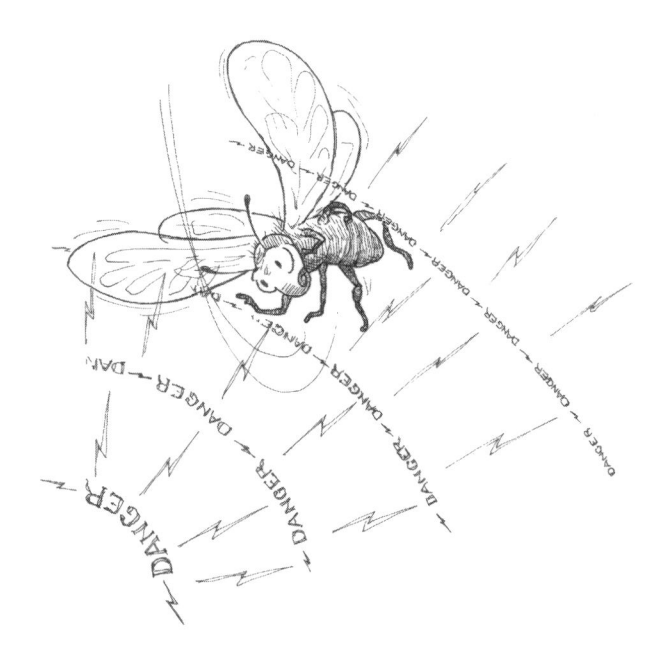

The invention of a way to detect aircraft through the use of radio waves that bounce off of a target and return to a receiving device was a technological achievement developed during World War II. It is known as radar (RAdio Detection And Ranging). A similar principal using sound waves was then applied to submarine detection, resulting in the development of sonar (SOund Navigation And Ranging). Sonar has progressed to the point where today even the public can possess devices like "fish finders" that help anglers locate schools of fish. Some of these are even sophisticated enough to indicate the size of the fish and the depth at which they are swimming.

Unsurprisingly, insects make use of the equivalent of both radar and sonar. Moths in the families **Erebidae** (tiger moths and kin), **Noctuidae** (owlet moths), and **Geometridae** (inchworm moths) can detect the chirps of **echo-locating** bats. When moths are exposed to bat calls, they instantly take evasive action to avoid being eaten. They fly in erratic loops, take sharp nosedives, or simply accelerate and fly away at top speed. Experiments have shown that moths rendered deaf because their "ears", or **tympana**, have been disabled cannot hear a bat approaching and are quickly consumed by the winged mammal.

Wait, it gets even more weird. Many moths are afflicted by the Moth Ear Mite, *Dicrocheles phalaenodectes* (**Mesostigmata: Laelapidae**). Invariably, the mites are found in only one of the two ears, located on each side of the anterior of the moth's abdomen. Because the mites pierce the membrane, it renders the moth deaf, but only in the one ear. The moth can still hear bats through the unaffected ear.

At least one species of green lacewing, *Chrysopa carnea*, has tympanal organs that are also used to avoid predation by echo-locating bats. Some mantids are likewise endowed with such organs, as are some scarab and tiger beetles. It is likely that we will discover the same capacity in still other insects.

Sound production in aquatic insects can be compared to sonar. Larvae of caddisflies, nymphs of dragonflies, immature and adult true bugs, and adult beetles are among those that can produce sound, or at least vibrations, in water. The sounds serve various functions, but the most common are defense and communication. Whirligig beetles (**Coleoptera: Gyrinidae**) are the bullet-shaped beetles you see swimming erratically on the surface of ponds, lakes, and slow-moving rivers. They are able to avoid colliding with each other by using the waves they generate while moving. They are also able to detect prey in this way. The waves the whirligigs generate bounce back to them, allowing them to detect the position of objects on the water surface. The returning waves can even inform the beetle if the object is another whirligig or a potential meal.

Radar-jamming Device

Radar is basically the emission of directional sound waves and the interpretation of their behavior upon their return to a receiver. The information is used to pinpoint the location, speed, and direction of an object, usually for the purpose of its destruction or apprehension. Take a speeding vehicle, for example. A "radar gun" is used to gauge the speed of traffic, and any motorist exceeding the speed limit is then pursued and ticketed by law enforcement. Some drivers may employ devices to jam those radar guns and escape a citation. The jammers usually work by emitting their own frequencies of sound that disrupt the radar signals.

Some insects use similar tactics to avoid predators, specifically bats that use echo-location to find and pursue insect prey. Many tiger moths, colorful insects in the subfamily **Arctiinae** (**Erebidae**), not only have hearing organs in the abdomen to detect the calls of approaching bats, but they also have sound-producing tympanal organs in the thorax that generate species-specific sound frequencies. *Bertholdia trigona*, a tiger moth found from Colorado to Central America, is capable of making much louder clicks than other moths. The clicking of the moth results in "range interference", which compromises the bat's ability to judge the distance from its moth target. This appears to be the only moth for which this tactic holds true.

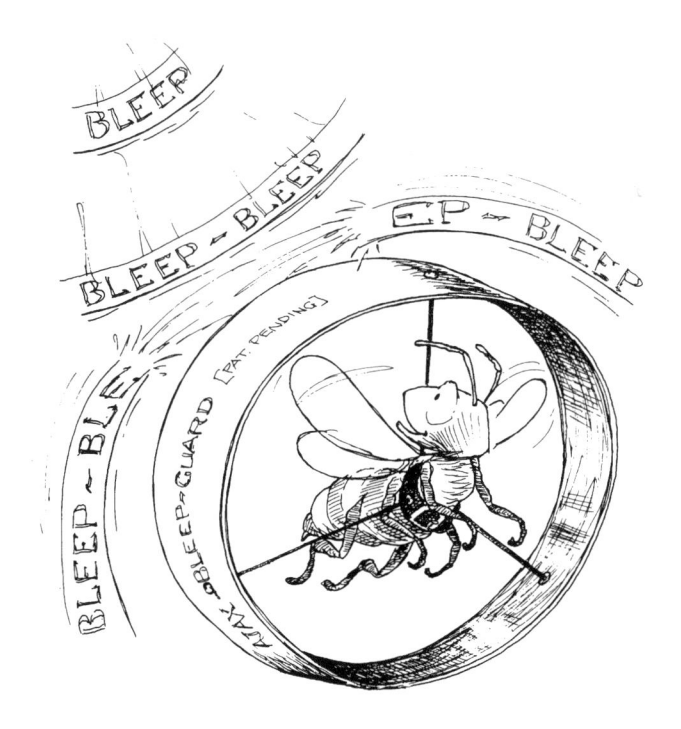

Meanwhile, the Dogbane Tiger Moth, *Cycnia tenera*, uses a different strategy. Its quieter clicks do not interfere with a bat's ultrasonic echolocation signals, but instead communicate to the predator that the moth is distasteful and therefore not worth pursuing. It is the auditory equivalent of warning colors. Most other tiger moths that produce sound are likely doing this rather than jamming bat echolocation.

Despite a variety of studies aimed at solving how moth sounds affect their predators, the phenomenon remains largely a mystery, and probably one with more than a single explanation.

Kamikaze, Suicide Bomber

Terrorism has become all too familiar a human tragedy that is often perpetrated by individuals willing to sacrifice their own lives in violent collisions or explosions designed for maximum casualties. It is a terrible yet effective means of inflicting devastation upon one's foes.

The Japanese term kamikaze, or "divine wind", first came into use during World War II. In a desperate act to win the war, Japanese pilots flew suicide missions by intentionally crashing their planes, laden with explosives, into the ships of Allied Forces. These pilots considered it an honor to die for their emperor and regarded their lives as a small price to pay for defense of their homeland.

Some social insects, in defense of their queen and vulnerable egg, larval, and pupal siblings, also engage in suicidal behavior. The most familiar example is the worker honey bee, genus *Apis*. This female soldier has barbed stingers that anchor into her enemies' flesh. As the bee departs, vital organs are torn from her body, dooming her to death, but her stinger remains embedded, and the attached venom gland continues pumping poison into the wound. An all-out assault by hundreds of worker bees can result in fatality for the sting victim, too.

Less well-known than the honey bees are the exploding soldier termites, *Globitermes sulphureus*, (**Termitidae**) of Southeast Asia. The insects possess an enormous frontal gland that takes up much of the abdominal and thoracic cavities. This gland manufactures a highly sticky defensive compound that is released by violent contractions at a weak point in the abdomen. The result is explosive dispersal of the sticky chemical over attacking predators, which are most often ants. This self-induced rupture of the body wall is known as **autothysis**.

Believe it or not, autothysis is also known in the carpenter ants *Camponotus cylindricus* of Borneo, and *C. saundersi* of Malaysia. Once again, these agents detonate a sticky goo along with highly irritating compounds that, because of their glue-like nature, cannot be wiped off by the victim. Larger, predatory ants are the chief targets of this defense tactic. The carpenter ants usually engage the enemy first with their mandibles, and then...BOOM!

Armor

Humans who engage in battle have employed the use of armor dating back at least to the "chain mail" of Medieval times. Today, our technology has advanced to lighter yet more durable materials like the Kevlar used in bullet-proof vests. The anatomy of our species, with its internal skeleton, leaves us much more vulnerable than your average insect.

Insects have an exoskeleton, which means that their skeletons or body frames are on the outside of their bodies. There are many advantages to this structure. There is much more surface area available for muscle attachment, which accounts for the strength and agility of insects that is so often out of proportion to their size. Also, the exoskeleton varies from highly elastic to highly rigid, depending on the life stage and requirements of the insect inside it. It is essentially built-in armor in many insects, especially beetles. The exoskeleton is articulated at many points, especially the legs, which explains why insects and their kin are known as arthropods, or "joint-footed" invertebrates.

Insects not endowed with a heavily sclerotized exoskeleton may employ other materials with which to protect their softer bodies. Caterpillars of bagworm moths (**Lepidoptera: Psychidae**) construct "mobile homes" of silk and twigs and other sturdy debris to deflect the attacks of predators. Caterpillars of casebearer moths in the family **Coleophoridae**, as well as some leaf beetle larvae in the family **Chrysomelidae** (subfamily **Cryptocephalinae**), encase themselves in dense capsules made from their own, dried feces. Still other leaf beetle larvae fashion a shield or umbrella of fecal material, which they thrust in the face of potential predators to discourage them from attacking.

Armored scale insects (**Hemiptera: Diaspididae**) secrete a hard-wax coating over their bodies that serves to protect them. In fact, most scale insects in the true bug superfamily **Coccoidea** do this. The secretions of the Indian Lac Insect, *Laccifer lacca* (**Hemiptera: Kerridae**) are harvested on farms in India and Thailand. The cultivation of the insects employs somewhere between one and three million rural people who help turn the waxy secretions into natural shellac.

Mimicry

They say that imitation is the sincerest form of flattery, but in the animal kingdom it is a matter of survival. Mimicry in its simplest form is appearing or behaving like another animal. Thus, children sometimes mimic their parents in dress and behavior, and some vocally-adept birds mimic sounds that they hear, including human conversation. However, insects take mimicry to the extreme.

There is no end to the mind-blowing examples of arthropod deception, but it takes two basic forms. **Batesian mimicry** (named after the English naturalist Henry Walter Bates) occurs when a palatable insect (the mimic) resembles a distasteful or dangerous "model" insect. Many kinds of flies, for example, emulate stinging wasps or bees in their colorful appearance and hovering behavior. **Müllerian mimicry**, named for German biologist Fritz Müller, is when several species of well-defended insects share a common color pattern. Müller observed this in butterflies in the Brazilian rainforest and then created mathematical models to reinforce his hypothesis of the mutual benefit of shared, bold color patterns.

The famous mimicry of the Monarch butterfly by the Viceroy butterfly started out as a classic example of Batesian mimicry, as the Monarch is toxic due to its caterpillar diet of milkweed, whereas the Viceroy was considered a tasty species by comparison. However, it was eventually proven that the Viceroy is not palatable, either, making this a case of Müllerian mimicry instead.

A third form of mimicry that is peculiar to other insects that live in ant nests, or at least interact with ants, is **Wasmanian mimicry.** The larvae of *Microdon* syrphid flies are slug-like creatures that prey on ant larvae with impunity in part because of chemicals in their integument that mimic those of their ant hosts.

Entire books can be, and have been, written about mimicry in the insect world, and we are learning of more instances all the time. Mimicry may be the most overwhelming roadblock to naturalists wishing to learn how to identify insects. Just when you thought it was a beetle, it turns out to be a moth.

Arms Race

An unfortunate recurring theme in world history is war. The constant quest for military superiority has led to an ever-escalating arms race in which newer, deadlier weapons are developed to answer the advances of adversaries. While human beings are driven by greed, power, revenge, and other motivations well within our control, insects evolve new physical and behavioral responses in order to survive. It is a case of adapt or perish.

At first, life forms were soft-bodied, but 500 to 600 million years ago during the Cambrian period of geological history, a relatively sudden transition took place during which many animals became heavily armored and spined. Although no insects are known to have existed at this time, these insect ancestors entered the "arms race" with a vengeance, and that competition continues today. Many insects (such as beetles) are heavily sclerotized and also have the ability to deliver nasty bites or stings to protect themselves from enemies.

Another arms race occurs between herbivorous insects and the plants they feed on. Plants with more complex defense systems, usually chemical defenses, have a greater diversity of herbivores that feed on them. The ecological term for these adaptations and counter-adaptations is coevolution. Insect digestive systems evolve to disarm certain plant toxins to allow them to exploit those plants. So milkweeds, which possess potent poisons called **cardiac glycocides**, are attacked by a suite of insects capable of not only overcoming those chemicals, but also sequestering the toxins for defense against their own predators.

Another interesting coevolutionary relationship exists between certain plants and ants. The plants, collectively called **myrmecophytes**, provide nesting sites for ants in return for their protection from herbivores like insects and browsing mammals. A few of these plants have even evolved elaborate structures to house the ant colonies. For example, *Myrmecodia*, ipiphytic plants found in Southeast Asia, form large, hollow pseudobulbs underground that are occupied by *Iridomyrmex* ants. Perhaps the best-known examples of ant-plant mutualism are the bullhorn acacia trees of the New World tropics. The hollow thorns are home to *Pseudomyrmex* ants that run off all herbivores. Meanwhile, the trees produce tiny, expendable **"beltian bodies"** to feed the ants. The miniscule growths are rich capsules of lipids (fats), proteins, and sugars, which are all the nutrients ants need to go from egg to adult.

Cannon

The best weapons, of course, are non-lethal and designed to inflict pain or punishment of a degree certain to stop unlawful behavior and deter repetition of that behavior in the future. One such device is the water cannon used to disperse participants in riots and other violent group conduct. The high pressure of the water cannon knocks people over, and they remain wet and uncomfortable for hours.

The original canon may be the explosive apparatus of the bombardier beetles, genus *Brachinus*. that expel a hot (212° F), pulsating spray of irritating quinones when molested. These quinones are so reactive that they originate as precursor chemical compounds in separate glands in the beetle's abdomen. They combine in a "reaction chamber", and the resulting compound is fired from the insect's anus. Upon contact with the air, the quinones produce an audible "pop" along with a visible cloud of vapor. The beetle can aim its cannon in any direction with great accuracy. A predator bent on making the insect a menu item will never again make that mistake after a blast from the beetle's anus.

A more lethal form of "flatulence" is a product of the larva of the beaded lacewings (**Neuroptera: Berothidae**). Laboratory observations of tiny larvae of *Lomamyia latipennis* have shown that what the insect lacks in the force and fireworks of a bombardier beetle, it makes up for in virulence. Beaded lacewing larvae are termite predators whose attention you do not want to attract inside a colony because the soldiers could cut you to pieces. So, the lacewing larva approaches harmless worker termites cautiously, waving its rear end in front of the termites' faces. It is expelling a "**vapor-phase toxicant**" that does not repel or otherwise alarm the termites, but knocks the average termite unconscious in one to three minutes. The lacewing larva is then free to consume its comatose victims.

Prisoners of War

One of the most demoralizing and intolerable aspects of warfare is the taking of prisoners, who are often tortured or enslaved as informants, labor, or sex slaves, among other cruel practices. Slavery of humans persists to this day in other contexts as well, even though it is widely regarded as a crime against humanity.

Strange as it may seem, slavery is common among ants, with at least 35 species recorded as depending on the slave labor of other ant species for the survival of their colonies. Indeed, most obligatory slave-making ant species are so specialized that they would starve if deprived of their slaves. The classic example of this is the genus *Polyergus*, known as the "Amazon ants". Worker Amazons have simple, sickle-shaped mandibles (jaws) perfectly adapted to pierce the armored exoskeletons of other ants, specifically those in the genus *Formica*. As a result of this specialized anatomy, Amazons cannot feed themselves.

A slave raid by a *Polyergus* colony is a spectacle to behold. After a scout has located a colony of *Formica*, it notifies its nestmates, and an attack ensues. The Amazons march in a regimented column straight to their destination along a pheromone trail the scout laid down to guide them. Upon arriving, the Amazon workers set about dispatching the *Formica* workers, driving their jaws into their victims' heads. They also emit a chemical called a "propaganda pheromone" that causes the other *Formica* workers to hastily abandon their underground nest. The Amazons are then free to raid the nest for larvae and pupae, which they carry back to their own colony. The *Formica* larvae, and some pupae, will be consumed as food, but the majority of pupa are allowed to mature into adult Formica ants that then serve the *Polyergus* colony by raising the queen Amazon's offspring to adulthood and feeding the adult workers, queen, and males.

Not all species of slave-making ants are obliged to live this lifestyle. Many are opportunists or "facultative slave-makers" that can enslave other colonies when an opportunity arises. Ironically, there are several facultative slave-makers in the genus *Formica*.

Slavery is not restricted to ants. Some social wasps display this behavior, too. The Parasitic Yellowjacket, *Dolichovespula arctica*, does not even have a worker caste. Instead, the queen infiltrates a young colony of another yellowjacket species (usually the Aerial Yellowjacket, *Dolichovespula arenaria*) and lives alongside the resident queen until the colony develops to a small corps of workers. The parasitic queen then assassinates or, rarely, simply evicts, the rightful queen and employs her workers to raise future parasitic queens and males.

References

Akre, R. D. 1982. Social wasps, pp. 1-105. *In* H. R. Herman [ed.], Social insects, Vol. 4. Academic, New York.

Atkins, M. D. 1980. Introduction to insect behavior. MacMillan, New York.

Bell, R. and R. T. Carde [eds.]. 1984. Chemical ecology of insects. Sinauer, Sunderland, MA.

Blum, M. S. 1981. Chemical defenses of arthropods. Academic, New York.

Borror, D. J., C. A. Triplehorn, and N. F. Johnson. 1989. An introduction to the study of insects. Saunders, Philadelphia, PA.

Brower, L. P. 1969. Ecological chemistry. Sci. Amer. 220: 22-29.

Corcoran, A. J., J. R. Barber, N. I. Hristov, and W. E. Conner. How do tiger moths jam bat sonar? 2011. J. Experimental Biology 214:2416-2425.

Corcoran, A. J., W. E. Conner, and J. R. Barber. 2010. Anti-bat tiger moth sounds: form and function. Current Zoology 56: 358-369.

Cott, H. B. 1966. Adaptive coloration in animals. Methuen, London.

Cowen, R. 1990. Parasite power. Sci. News 138: 200-202.

Dunning, D. C. and K. D. Roeder. 1965. Moth sounds and the insect-catching behavior of bats. Science 147: 173-174.

Eisner, T., E. Van Tassell, and J. E. Carrel. 1967. Defensive use of a fecal shield by a beetle larva. Science 158: 1471-1473.

Erlich, P. R. and P. H. Raven. 1967. Butterflies and plants. Sci. Amer. 216: 104-113.

Howard, R. W., R. D. Akre, and W. B. Garnett. 1990. Chemical mimicry in an obligate predator of carpenter ants (Hymenoptera: Formicidae). Ann. Ent. Soc. Amer. 83: 607-616.

Johnson, J. and K. Hagen. 1818. A neuropterous larva uses an allomone to attack termites. Nature 289: 506-507.

Kettlewell, H. B. D. 1965. Insect survival and selection for pattern. Science 148: 1290-1296.

Kettlewell, H. B. D. 1973. The evolution of melanism. Clarendon, New York.

Kirk, V. M. and B. J. Dupraz. 1972. Discharge by a female ground beetle, *Pterostichus lucublandus* (Coleoptera: Carabidae), used as a defense against males. Ann. Ent. Soc. Am. 65: 513.

Maschwitz, U. and E. Maschwitz. 1974. Platzende arbeiterinnen: Eine neue art der Feindabwehr bei sozialen hautfluglern. Oecologia 14: 289-294.

Metcalf, C. L., W. P. Flint, and R. L. Metcalf. 1962. Destructive and useful insects, 4th ed. McGraw-Hill, New York.

Miller, L.A. and E.G. MacLeod. 1966. Ultrasonic sensitivity: a tympanal receptor in the green lacewing *Chrysopa carnea*. Science 154: 891-893.

Miller, L.A. and A. Surlyke. 2001. How some insects detect and avoid being eaten by bats: tactics and countertactics of prey and predator. Biosci. 51: 570-581.

Papageorgis, C. 1975. Mimicry in Neotropical butterflies. Amer. Sci. 63: 522-532.

Pasteur, G. 1982. A classificatory review of mimicry systems. Ann. Rev. Ecol. Syst. 13: 169-199.

Rettenmeyer, C.W. 1970. Insect mimicry. Ann. Rev. Ent. 15: 43-74.

Ritland, D. B. and L. P. Brower. 1991. The viceroy butterfly is not a Batesian mimic. Nature 350: 497-498.

Roeder, K. D. 1965. Moths and ultrasound. Sci. Amer. 212: 94-102.

Roeder, K. D. and A. E. Treat. 1961. The detection and evasion of bats by moths. Amer. Sci. 49: 135-148.

Roth, L. M. and T. Eisner. 1962. Chemical defenses of arthropods. Ann. Rev. Ent. 7: 107-136.

Sargent, T. D. 1966. Background selections of geometrid and noctuid moths. Science 154: 1674-1675.

Schildknecht, H. 1970. The defensive chemistry of land and water beetles. Angew. Chem. 9: 1-9.

Schildknecht, H. 1971. Evolutionary peaks in the defensive chemistry of insects. Endeavour 30: 136-141.

Suomi, D. 1988. Snailcase bagworm. Cooperative Extension, Wash. State Univ. Ext. Bull. 1485. Topoff, H. 1984. Invasion of the body snatchers. Nat. Hist. 93: 78-84.

Tucker, V. A. 1969. Wave making by whirligig beetles (Gyrinidae). Science 166: 897-899.

Wilson, E. O. 1971. The insect societies. Belknap/Harvard Univ. Press, Cambridge, MA.

Wilson, E. O. 1975. Slavery in ants. Sci. Amer. 232: 32-36.

Miscellaneous Categories

Vampire

An enduring myth and staple of horror films is the idea of vampires. There are no humans that feed on the blood of other humans, let alone that are immortal unless exposed to sunlight or staked through the heart. However, there are definitely other creatures that will consume human blood, including three species of vampire bats.

A surprising array of insects are blood-feeders, though human beings are usually a "host of last resort". Exceptions include the Bed Bug and various sucking lice including the Head Louse, Body Louse, and Crab Louse. These insects occur nowhere else but on *Homo sapiens*. We battle plenty of other diminutive vampires, though: mosquitoes, black flies, horse flies, deer flies, stable flies, tsetse flies, no-see-ums (biting midges), sand flies, bloodsucking conenose assassin bugs ("kissing bugs"), and fleas. The larvae of the Congo Floor Maggot, *Auchmeromyia luteola*, (**Diptera: Calliphoridae**) also feeds on the blood of people and other animals in Africa. Believe it or not, there are even "vampire moths" in the genus *Calyptra* (**Lepidoptera: Erebidae**), the males of which are known to feed on the blood of mammals, including people on rare occasions. One could add various mites and ticks to the list, except they are not insects. They are arachnids, allied to spiders.

Not surprisingly, there are also insects (and mites) that feed on the blood of....insects. Chief among them are most of the biting midges in the family **Ceratopogonidae**. These tiny flies will puncture a wing vein or intersegmental membrane to reach their victim's hemolymph. Since insect blood does not flow through vessels like ours does, any portal into the body cavity will allow a blood-feeding insect to feast on its insect host.

Perhaps the most bizarre example of insects that feed on the blood of other insects are the so-called "Dracula Ants", *Adetomyrma venatrix*, found in Madagascar. This primitive social species forms colonies of as many as 10,000 workers. They hunt other insects by stinging them to death or paralysis for transport back to the nest to feed the colony's larvae. This is where it gets weird. The 1/16th inch-long adult ants regularly pierce the cuticle of their baby sisters in order to drink their siblings' hemolymph. It is this behavior that made headlines, as well as the facts that the species was unknown to science until 1994 and is found nowhere but on Madagascar.

Polluters

Mankind certainly excels at generating pollutants, from our own human waste and the garbage we manufacture as a result of over-packaging to the toxins released into the air and water during natural resource extraction and the burning of fossil fuels and wood. Pesticides and fertilizers are washed into rivers, streams, lakes, and oceans. There seems to be no end to it. There are even enormous patches of plastic floating in our oceans and suffocating coral reefs.

Insects pollute, too, but generally the magnitude of their impact pales compared to ours. It has been widely cited that termite.... "flatulence" is one of the leading sources of atmospheric methane gas, which contributes to global warming as it breaks down into carbon dioxide, ozone, and water, all of which absorb heat. Some estimates put termite farts at up to fifteen percent of the total methane in our atmosphere, but that figure is debatable. It is more likely that the annual contribution of termites to methane production, while substantial, amounts to less than 15 Tg (teragrams or "tonnes") per year, or less than 5 percent of the total annual methane emissions.

Meanwhile, aquatic insects pollute, too. All insects must excrete nitrogen waste accumulated as products of protein metabolism. Most terrestrial insects convert these compounds into an insoluble and therefore non-toxic product called uric acid before they are excreted. Aquatic insects, however, discharge ammonia, a highly poisonous material, directly into the water. Since these insects live in an aquatic habitat, they are constantly awash in water, a cleansing solvent that quickly dilutes and disperses this toxic ammonia waste. Converting metabolic wastes into uric acid is an expensive (energy-consuming) process, so aquatic insects have an advantage over their terrestrial counterparts.

Gravediggers

The undertaker, mortician, embalmer, or funeral director, whichever name you choose, is a highly underappreciated occupation and an extremely important one at that. Proper disposal of deceased bodies is vital to public sanitation. A degree of compassion and reverence is also required, of course.

The natural world employs all manner of organisms to assist and complete the process of decomposition: turning dead animals and plants back into nutrients to nourish those individuals still living. No ceremony is needed; it is all very matter-of-fact. Insects are vital agents in this cycle, and one of them is even a gravedigger.

The burying or "sexton" beetles in the genus *Nicrophorus* (**Silphidae**) are experts at interring the bodies of small animals such as rodents, moles, shrews, birds, and even reptiles. They are fairly large, muscular insects that are cloaked in black with red, orange, or yellow accents, although some species lack those colorful markings. They may work in teams, male and female together, or alone (female only).

Flying in search of a corpse, a beetle will land when it detects one and crawl under the carcass to begin excavating. If the body is in a situation unsuitable for burial, the beetle turns upside down and, sliding along on its back while treading on the decedent, literally walks it to a proper resting place. Once a suitable resting place is located, the beetle sinks the body into the soil by digging under it, eventually forming a shallow grave. A female then turns it into a crypt by building a sturdy-walled chamber around the prize. Next, she strips fur or feathers from the carcass and forms it into a "meatball". After laying a small clutch of eggs in a crater atop the mass, she licks both the meatball and her eggs to coat them with antibacterial and antifungal substances that retard rotting and contamination. She does this continually after the eggs hatch. The larvae that emerge from the eggs are fed morsels by their doting mother until they are able to begin feeding on their own. Once they finish eating and growing, the larvae enter the pupa stage and complete their development to adulthood, eventually digging their way to freedom to start the cycle anew.

Food Storage

We store food for lean times and stock our pantries regularly with various ingredients for recipes. We can fruit and vegetables, freeze meats and other foods, and otherwise create a larder that we can pick from as need or want arises. Such foresight in provisioning is not unique to our species. Even insects do so, and they have created some ingenious solutions to food storage in the process.

Honeypot ants in the genus *Myrmecocystus* are native to the southwestern U.S. They earn their name from specialized worker ants in the colony called "repletes". The repletes have amazingly elastic membranes capable of accommodating large amounts of honeydew gathered from aphids, scales, and other insects. Foraging workers harvest the sweet liquid and return to the colony to feed the excess to the repletes. These living honey jars have abdomens that can expand to the size of a pea or even a grape when filled to capacity. Once filled, the repletes then act as living dispensers of liquid food when colony members beg from them. Native Americans knew of this phenomenon and dug deep underground to reach the sweet repletes for their own consumption.

Honey bees store food, too, of course, that is likewise exploited by humans who crave the sweet substance. Many wax cells, arranged in combs, are devoted to the storage of honey to benefit the colony during times when nectar sources are few and far between. Most other social bees, including bumble and stingless bees, store honey in wax vessels.

Surprisingly, some social wasps also store honey. The Honey Wasp, *Brachygastra mellifica*, is a prime example. This is a tropical species that ranges as far north as southern Texas and Arizona. Look inside the open combs of paper wasps (*Polistes*) and you may also see drops of honey inside the individual cells. The droplets offer an energy boost to worker wasps and fuel their activities of foraging and caring for their larval siblings.

Harvester ants of various genera, but particularly *Pogonomyrmex*, collect and store seeds in underground nest chambers. The cached morsels help sustain the colony during the winter, when other food sources such as dead or dying insects are in short supply. Other insects benefit from this practice, too. Many species of darkling beetles (**Coleoptera: Tenebrionidae**) can be seen hanging around the periphery of harvester ant mounds, feasting on the chaff and seeds littering the entrance.

Hypodermic

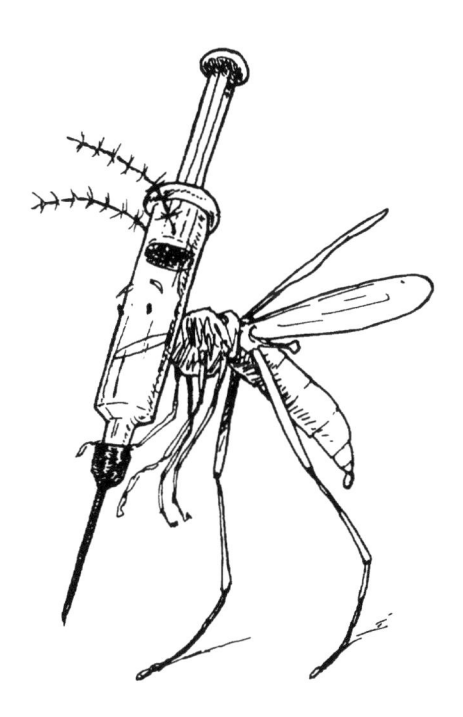

The use of hypodermic syringes and needles for the administration of medicines or for drawing blood is a common practice, as is, unfortunately, the use of the hypodermic syringe to "shoot up" illegal drugs. The syringe is armed with a needle, a sharply pointed, fine-caliber tube, to pierce the skin. The needle is used with a syringe that works on the principle of differential pressures between the blood vessels and the syringe. The original hypodermic, however, belongs to the insects. The mouth parts of most bloodsucking and plantsucking insects are sharply pointed, elongated structures that interlock to form a piercing tube. Food is carried up the mouth parts as the result of suction created by the expansion of diaphragm-like pumps in the insect's head. Such hypodermics can be found in many bloodsucking insects, including bed bugs (**Hemiptera: Cimicidae**), mosquitoes (**Diptera: Culicidae**), black flies (**Diptera: Simuliidae**), and stable flies (**Diptera: Muscidae**). All of these insects inject materials to inhibit blood clotting as they feed. Similar hypodermic-like mouth parts occur among aphids (**Aphididae**), scale insects (**Coccidae**), and true bugs (**Hemiptera**).

Antibiotics/Medicines

Insects, like humans, have trouble with microbes. Diving water beetles (**Coleoptera: Dytiscidae**) produce an antibiotic paste to keep their body surface microorganism-free. This helps keep the surface smooth for friction-free swimming and reduces the number of invasive bacteria with which they must contend. The components of the paste are benzoic acid and several phenols, particularly methyl p-hydroxybenzoate and p-hydroxybenz-aldehyde, plus a glycoprotein. The paste is excreted from special glands and spread onto their bodies with their hind legs while the beetles are sitting on emergent vegetation. The paste hardens, killing and encapsulating many microbes, and is washed off when the beetle returns to the water. These materials, particularly the methyl p-hydroxybenzoate and similar compounds, are used by humans to eliminate microbes from canned foods.

Humans have learned to exploit insect antibiotics for medicinal purposes. People have attributed curative powers to insects for hundreds of years. While in the 1600s these powers were ascribed to nearly every insect it is now known that certain insects do indeed have special medicinal properties. Perhaps the most famous case was recorded by Dr. W. S. Baer. While observing the wounds of soldiers in World War I, he noticed that injuries infested with greenbottle fly maggots did not develop gangrene whereas those treated and sometimes even dressed promptly frequently did. Further investigations showed that the maggots ate only putrid flesh, and that they were much more efficient than physicians in the cleaning of deep wounds. Because of this blow flies were sometimes reared under sterile conditions specifically to be used for this purpose, and in rare cases are even used today. An additional benefit was discovered in 1935 when scientists discovered that maggots secrete allantoin, a substance that promotes healing. Of course, mostly synthetic allantoin is used today.

Another medicine that insects produce is cantharidin, the defensive chemical that blister beetles release. A particularly well-known insect in this group is *Lytta vesicatoria*, the Spanish fly. Humans used extracts from this beetle as an aphrodisiac until we realized that this material is extremely harmful. It is still judiciously used today for the treatment of diseases of the urogenital tract and for the breeding of animals.

The venom from the sting of honey bees is probably the most widespread insect "medicine" in the world. Bee stings are used by many people to treat their arthritis and many other ailments. Royal jelly, another honey bee product, is also used by many, although it has no known medicinal value.

Recycling

It has become common practice in western civilizations to recycle materials as varied as plastics, aluminum and other metals, paper and cardboard, and glass. This is done as a measure of conserving these resources as well as limiting waste and curbing littering. There are few strict mandates, however, and one must look to the World War I and II eras for examples of stringent recycling requirements.

Other animals re-use materials, too, especially those that their body produces. Honey bees will re-use beeswax, a product of glands in the abdomen of worker bees. It is expensive in terms of energy to produce wax, so it is a valued commodity in each colony. The entire nest, which is composed of combs of hexagonal cells, is made of wax generated by the bees. Apiculturists (beekeepers) furnish their hives with artificial "foundations" inside removable frames on which the worker bees can build. These starter kits require less wax than if they built a comb from scratch.

Most social wasps, especially yellowjackets (*Vespula* and *Dolichovespula*), construct large nests entirely of paper. They manufacture their building material by harvesting wood and plant fibers and chewing them into a pulp that is then applied in thin strips. Each colorful stripe you see on the exterior of a paper nest represents one mouthful of pulp. As the nest grows, the interior layers, called envelope, of the nest surrounding the paper combs are torn down and then re-applied to expand the combs, strengthen the suspension columns between combs, or to add to the envelope's exterior layers. The nest thus expands from the inside out.

Spiders are not insects, but as fellow arthropods they, too, practice recycling. Orbweaver spiders are known to eat the remains of damaged and worn webs, recycling the silk for use in the production of a new web. Silk-producing insects probably do not do this; their use of silk is often for shelter and to protect the pupa stage, so these situations represent a "one-time only" use of silk.

When insects molt their rigid exoskeletons to grow, they leave behind an empty "husk" that many insect larvae and nymphs will consume. This serves two purposes: it returns to the insect some of those nutrients needed to build and maintain the new exoskeleton, and it eliminates evidence of the insect's presence. Many sharp-eyed predators like birds will notice a shed exoskeleton and then look for the insect that left it behind.

Problems with Insects

Insects are man's leading competitors for food and also pester us directly. They can afflict our livestock and pets, the structure of our homes and buildings, and infect us with diseases. There is scarcely any aspect of our lives that is immune to problems from insects.

One may take small comfort from the fact that insects are also plagued by other insects, even beyond the roles of prey and parasite victim. So goes this familiar excerpt from a Jonathan Swift poem, *On Poetry: A Rhapsody* (1733):

So, naturalists observe, a flea
Has smaller fleas that on him prey;
And these have smaller still to bite 'em
And so proceed *ad infinitum*.

Indeed, the biting midges (**Ceratopogonidae**) in the genus *Forcipomyia* are blood-sucking nuisances of mantids, walkingsticks, dragonflies, alderflies, lacewings, beetles, moths, crane flies, and even mosquitoes.

Some insects, plus other arthropods like mites and pseudoscorpions, will use larger insects as a means of transport from one location to another. This hitchhiking behavior is known as phoresy. One example is a tiny parasitoid wasp in the family **Scelionidae**. This wasp will ride on a female grasshopper and get off after the grasshopper lays her eggs. The wasp then lays her eggs in the grasshopper's egg pod.

Perhaps the most unlikely example of phoresy revolves around the Human Bot Fly, *Dermatobia hominis*, (**Diptera: Oestridae**) a bee-sized insect that develops as a larva under the skin of a human victim. The adult female fly does not lay her eggs directly on the host. Oh, no, that would be too simple. Instead, she accosts a female mosquito in midair, adheres her eggs beneath the mosquito's body, and flies off. When the female mosquito alights on a person, the body heat of the human triggers the bot fly eggs to hatch. Each tiny maggot then wriggles into a pore or follicle or into the wound left when the mosquito departs. They feed on subcutaneous tissue, often quite painfully, and grow to an uncomfortable size before eventually erupting and dropping to the ground, where they enter the pupa stage.

Some bees and wasps are disfigured by their own equivalent of a bot fly. Bizarre, tiny insects called twisted-winged parasites (**Strepsiptera: Stylopidae**) wedge themselves in between the abdominal segments of their host, distorting the body of their victim in the process. The wasp or bee picks up the stylopid when it visits a flower crawling with the parasite's larval stage. The larva or larvae burrows through the cuticle of the bee or wasp and lives as an internal parasite through up to seven molts before making a partial exit as a pupa. The capsule-like pupal case protruding between the host's abdominal segments is the only outward indication of a problem. Male stylopids are winged, and they emerge from the pupa to fly off in search of a female. Females emit a pheromone to attract males. The fertilized eggs of the female hatch inside her, and the tiny larvae emerge to begin the life cycle anew. Other members of the order Strepsiptera are parasites of other insects. [see also... "Beggars", other chapters]

Opposable Thumb

One of the defining attributes of primates, especially the great apes, of which *Homo sapiens* is one, is the opposable thumb. This wonder of anatomy allows us to grasp objects easily and manipulate them with a degree of skill that most other animals cannot achieve. Alas, we cannot even claim that as something that sets us apart from non-primates. The Giant Panda has a fixed digit with which it can seize and manipulate bamboo stalks as it chews on them; but even some insects possess similar features, even if one has to bend the rules a little.

Sucking lice (**Psocodea**: superfamily **Anoplura**) can be considered to have an opposable thumb if a little "poetic license" is permitted. The tarsi ("feet") of sucking lice are one-segmented and have a single, large claw that fits against a thumb-like knob at the end of their lower leg (tibia). This combination of an opposable thumb and claw is used to cling to the hairs

127

of the host mammal. Indeed, all the scratching and nipping of the host rarely succeeds in dislodging a louse, so sure is its grip.

A similar arrangement of claw and leg segments is found in small parasitic wasps in the family **Dryinidae**. The females of most have the front tarsi modified into a "chela", a pincer-like structure akin to a crab's claw. In dryinid wasps, this consists of an enlarged but slender, blade-like claw and the often-serrated tarsal segment to which it is attached. This scissor-like or clamp-like device is quite flexible (in contrast to the louse) and used to grab the host so the wasp can lay an egg on it. Her larval offspring will then feed as an external parasite. Most dryinid hosts are leafhoppers and related true bugs.

Predatory insects like mantids, mantidflies, ambush bugs, assassin bugs, and some kind of true flies, to name but a few, also have modifications to their leg anatomy that help them seize prey in a vise-like death grip. In most cases, the tibia segment of the front leg is blade-like and fits tightly into grooves along the very muscular femur segment behind it. The underside of the femur is often studded with spines or teeth for extra purchase in securing a struggling victim so that the tibia and femur act together as opposing forces for grip strength and manipulation of prey.

The Wheel

The invention of wheels is considered one of the greatest accomplishments in human history. Their creation dates back to at least 3500 BCE, the time when potter's wheels came into use in Mesopotamia. It would be another 300 years before they were applied to transportation, on chariots.

It may be stretching the truth only slightly to say that insects were the first organisms to employ the wheel as a means of locomotion. The larva of the Southeastern Beach Tiger Beetle, *Cicindela dorsalis media*, (**Carabidae**) normally lives in a vertical shaft it tunnels into the sand. When it wants or needs to escape, it does a remarkable thing. It leaps out of its burrow, grabs its rear-most body segments in its jaws, and assumes a circular configuration while still airborne from its leap. The wind catches it and propels it over the beach ahead of any would-be predator or parasite or high tide. Wind speed and the topography of the sand surface have a lot to do with the speed and success of such excursions, but a larva can travel upwards of 200 feet at a time in this manner.

The tiger beetle larva is not alone in this behavior. The caterpillar of the Mother of Pearl Moth, *Pleuroptya ruralis*, (**Crambidae**) aka the "Bean Webworm", can also form a wheel by gripping its rear end in its mouth and rolling away from danger. The caterpillar of the Carnation Tortrix, *Cacoecimorpha pronubana*, (**Tortricidae**) likewise uses the "recoil-and-roll" method of fast escape.

Although spiders are only distantly related to insects, it is worth noting that the "wheel spiders" of the genus *Carparachne*, (**Araneae: Sparassidae**) found in southern Africa, will fold their legs into a posture that allows them to cartwheel down sand dunes on their "knees" to escape predators like spider wasps that dig the spiders out of their burrows.

Meanwhile, mating dragonflies and damselflies join together to form the heart-shaped "wheel position" to accomplish insemination. The male grasps the female by her "neck" using claspers at the tip of his abdomen. She in turn stretches her abdomen forward and beneath his to reach the reproductive organs at the front of his abdomen. This unique and seemingly awkward position does not hinder the pair in the least, and they fly perfectly well while in tandem.

Mechanical Gears

At the risk of readers believing that the authors are now just "making stuff up", we offer a recent discovery of mechanical gears in an insect that, under scanning electron microscope magnification, look exactly like what you might expect in a piece of human-engineered machinery.

It turns out that nymphs (juvenile life stages) of the planthopper *Issus coleoptratus* (**Hemiptera: Issidae**) each possess modifications to their anatomy that look and act as gears to perfectly synchronize the kicks of their hind legs as the insect makes leaping maneuvers to escape danger. Whereas the human invention of gears dates back to about 300 BCE and is attributed to Greek engineers in the city of Alexandria, it took until September of 2013 for scientists to publish the discovery of the insect gears. In fairness, the bug is under 1/16th of an inch long, and its itty-bitty "engine" only occupies the basal segment of each hind leg. Interestingly, the adult insect lacks the gears and instead relies on a different mechanism for achieving its jumps. This alternate strategy may have resulted from the fact that the adult insect does not molt and therefore has no way to regenerate damaged teeth on the gears.

Ironically, adults of the large assassin bug known as the Wheel Bug, *Arilus cristatus*, appear to have a large cog embedded in the top of their thorax that creates an ornamental crest that serves no obvious function but is highly conspicuous. Maybe that cog is a reminder to potential predators that the Wheel Bug can defend itself with an excruciating bite from its beak-like mouthparts.

Velcro

Velcro was developed in 1948 by a Swiss engineer, George de Mestral, who was curious why cockleburs stuck to his socks and to his dog's fur after a walk in the woods. He examined the spiny seed pods with the aid of a microscope and discovered they were covered with hundreds of tiny hooks that attached themselves to anything loopy, like tangled hair or fluffy fabrics. He invented a method of duplicating the hook and loop configuration with nylon. He dubbed the product "velcro", a combination of the French words for velvet (velour) and hook (crochet). The basic patent expired in 1978, and dozens of manufacturers throughout the world currently produce variations of the concept, applying it to every manner of product that requires fasteners or fastening.

Not only plants, but also insects have arrived at the same type of morphology that links hooked spines and rough surfaces. Larvae of many ant genera, including *Pheidole, Crematogaster, Cephalotes, Temnothorax*, and *Strumigenys*, have hooked "anchor hairs" that allow the worker ants tending them to literally hang their immature siblings on the nursery chamber walls and/or ceilings. This allows for maximum chamber occupancy and also permits the adult worker ants to transport several larvae at a time if they need to be moved to another room or evacuated in the event of a disaster. Interestingly, ant larvae at different stages of development may not have these anchor hairs. The efficacy of these anchor hooks was proven through experiments in which some larval ants were given "haircuts" and could no longer be suspended from vertical surfaces after they were trimmed.

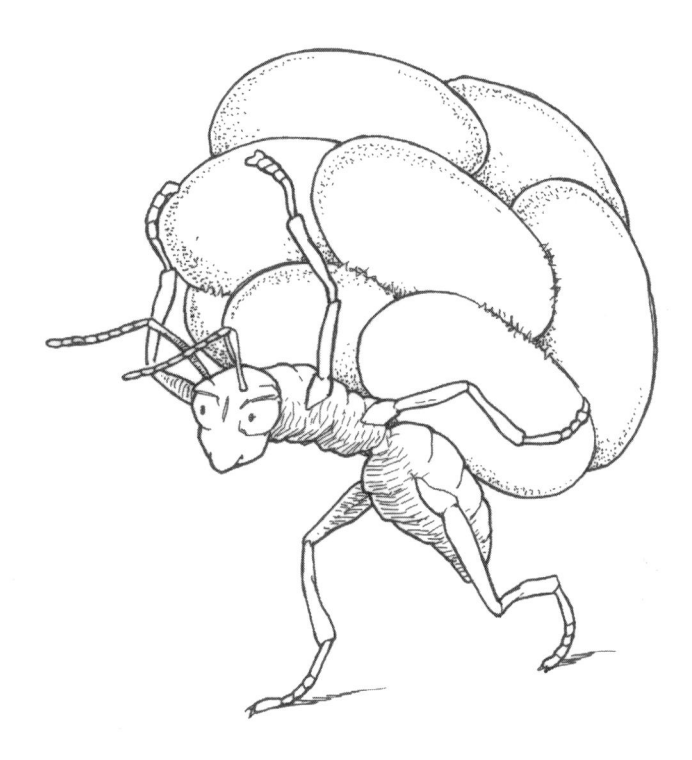

Larvae of the weevil *Eucoeliodes mirabilis*, found on certain plants in central and southern Europe, employ a shield of their own fecal matter that they carry atop the length of their bodies for protection from predators. The dorsal (top) surface of each larva is covered in a dense carpet of very short hairs called microtrichia. These setae are hooked at the tip and anchor the fecal shield over the insect so that the waste material never touches the larva's body. A different arrangement of the microtrichia around the larva's spiracles allow the immature insect to breathe freely, without obstruction from pieces of the fecal shield.

In Brazil, workers of the rainforest ant *Azteca andreae* wait in ambush along the leaf edges of their host tree, *Cecropia obtusa*. The underside of the leaves are hairy, with tangled loops like half of a velcro fastener. The ants have hooked claws on their feet, so when a larger insect lands on the leaf intent on getting a vegetarian meal, the ants bite it and hold fast with their claws anchored in the leaf's fluffy undersurface. Huge insects like grasshoppers can thus become the prey of these miniscule ants.

Baskets and Pots

Baskets and pots were among the first human inventions for transporting and storing objects. We know definitively that twined baskets were used as far back as 7000 BCE in Oasisamerica (part of today's southwestern U.S. that extends from western Utah westward and southeast through Baja California and the southern reaches of the Mexican state of Chihuahua). Woven baskets probably originated in the Middle East around 8000 BCE, but the jury is still out since objects crafted from plant materials do not always endure. Clay pottery had its beginnings in China and Japan, around 14,000 BCE.

The most familiar equivalent of baskets in the insect world are the "pollen" baskets on the hind legs of female (queen and worker) bees. Pollen baskets are a physical adaptation of the bee's hind tibia and foretarsal segments. They are expanded and flattened, with a concave exterior surface fringed with long, curved hairs. The technical term for the pollen basket is the "**corbicula**". Pollen baskets are designed to carry sticky wads of pollen mixed with small amounts of nectar. This specialized anatomy seems confined almost exclusively to social bees like honey bees, bumble bees, and stingless bees, with only a few solitary bee genera exhibiting similar morphology.

Solitary bees typically employ a different pollen-carrying strategy. Most carry dry pollen in thick brushes of long hairs that form a "**scopa**" on various parts of the hind leg or, in the case of leafcutter, mason, woolcarder, and resin bees, on the underside of the abdomen.

Many insects, especially certain solitary bees and wasps, are expert masons, as discussed under "Mud and Masonry" in the "Architecture and Engineering" section of this book. Besides the potter wasps in the genus *Eumenes*, there are other genera in the Eumeninae subfamily of vespid wasps that craft lovely examples of pottery that would rival at least a novice human ceramics maker. *Delta*, found mostly in tropical Africa and Asia; *Phimenes*, which occur in Malaysia, Indonesia, and tropical Australia; and *Zeta*, of the New World tropics, are among the true potter wasps, all of which belong to the tribe Eumenini. All use their "pots" as storage vessels for paralyzed caterpillars to be used as food for a single larval wasp offspring per pot.

Flypaper

In August of 1926, Jersey City's Olga Berghorn applied for a U.S. patent for "sticky fly paper", which she designed to attract and trap house flies and other filth flies that plague people and livestock indoors. Flypaper, which remains an important weapon against these pest and nuisance insects, now consists of a ribbon of yellow paper impregnated with a very tacky adhesive that effectively traps any flying insect that lands on it. Flypaper is typically suspended from a ceiling, pipe, or other overhead foundation and is simply discarded once its surface is covered in dead dipterans.

Guess what? Other insects were onto this idea, at least in the evolutionary sense, long before humans. Certain assassin bugs, particularly those in the common genus *Zelus*, are literally "armed" with flypaper. These slender true bugs prey on other insects using their front and sometimes middle legs to secure their struggling snack. Although they are not particularly muscular insects and each pair of legs seems unmodified for the task of dispatching another living organism, the fore and middle tibia segments of their legs bear a secret weapon: a dense coating of "sundew hairs". Like the leaves of the carnivorous sundew plant, the legs of the *Zelus* assassin bug boast an array of hairs connected to glands that secrete a supremely tacky substance. Once in the grip of one of these bugs, a victim has little chance of escape before the assassin delivers a paralyzing bite with its beak-like rostrum. Game over.

GPS

Today's standard of human navigation is the Global Positioning System (GPS). Our digital devices, which are the butt of an endless stream of jokes relating to our overreliance on them, depend on satellite technology to pinpoint our present location and desired destination. Insects don't need no stinking satellites; they have their own built-in version of GPS. Maybe several systems, in fact. We discuss the different types of compass mechanisms insects employ in the Transportation chapter of this book, but there is more to the story.

One of the most important and amazing means of "bug nav" is the ability of insects to perceive and utilize polarized light. Light waves from the sun, when they enter the Earth's atmosphere, vibrate in various planes at right angles to their path of travel. As the light waves collide with minute dust particles and gas molecules, the vibrations eventually occur in only one plane. The light waves are considered to be "plane polarized", and it is this polarization that insects frequently use to navigate.

Interestingly, insects seem to only use polarized light for straight-line travel when polarized light is at the zenith: directly overhead. In midsummer, this zenith of polarized light, arcing from north to south, occurs at sunrise. At the approach of midday, the band of polarized light nearly vanishes into the northern horizon, and insects either cease to travel, or will fly lower and around obstacles instead of over them. This change in behavior also occurs when clouds intrude over the polarized light skyscape. So, the lack of insect flight activity around noon in summer has less to do with heat than a lack of polarized light for them to steer by.

Researchers studying nocturnal "tumblebug" dung-rolling scarab beetles in Africa recently discovered how these seemingly clumsy insects can manage to stay on course to the destination where they want to bury their sphere of manure. The insect frequently pauses to climb atop the ball, then turns in circles before resuming its laborious ball-pushing. During those "dances" it is taking snapshots of the night sky, noting the positions of the moon, stars, and polarized sunlight (reflected off the moon). It archives those images in its brain and then matches them against the sky as it progressively moves its ball. Meanwhile, where did we park the car?

References

Borror, D. J., C. A. Triplehorn, and N. F. Johnson. 1989. An introduction to the study of insects. Saunders, Philadelphia, PA.

Borror, D. J. and R. E. White. 1970. A field guide to the insects of America North of Mexico. Houghton Mifflin, Boston, MA.

Burrows, M. and G. Sutton. 2013. Interacting gears synchronize propulsive leg movements in a jumping insect. Science 341: 1254 – 1256.

Fraser, P. J., R.A. Rasmussen, J.W. Creffield, J.R. French and M.A.K. Khalil. 1986. Termites and global methane – another assessment. J. Atmospheric Chem. 4: 295 – 310.

Glancey, M., C. E. Stringer, Jr., C. H. Craig, P. M. Bishop, and B. B. Martin. 1973. Evidence of a replete caste in the fire ant, *Solenopsis invicta*. Ann. Ent. Soc. Amer. 66: 233-234.

Harwood, R. F. and M. T. James. 1979. Entomology in human and animal health. Macmillian, New York.

Lofgren, C. S. and R. K. Vander Meer [eds.]. 1986. Fire ants and leaf-cutting ants. Westview, Boulder, CO.

Oldroyd, H. 1964. The natural history of flies. Norton, New York.

Philip, C. B. 1931. The Tabanidae (horse flies) of Minnesota, with special reference to their biologies and taxonomy. Minn. Agr. Exp. Sta. Tech. Bull. 80: 1-132.

Richards, O. W. 1978. The social wasps of the Americas excluding the Vespinae. British Museum (Nat. Hist.), London.

Romoser, W.S. 1981. The science of entomology. Macmillan, New York.

Sabrosky, C. W., G. F. Bennett, and T. L. Whitworth. 1989. Bird blow flies (*Protocalliphora*) in North America (Diptera: Calliphoridae) with notes on the Palearctic species. Smithson. Inst. Press, Washington, D. C.

Strassman, J. E. 1979. Honey caches help female paper wasps (*Polistes annularis*) survive Texas winters. Science 204: 207-209.

Wellington, W. G. 1974. "A Special Light to Steer By", *Nat. Hist.* LXXXIII: 10. 47-52.

Wheeler, W. M. 1910. Ants: Their structure, development and behavior. Columbia Univ. Press, New York.

Epilogue

We have seen how many aspects of insect biology correspond to human culture, but humanity owes much of its continuing success to insects in other ways. Insects have inspired our art, from poetry to paintings to film. Many of our inventions can be traced back to keen observations of insects. We look to insects as models for improved flight capabilities, and in robotics. We even seek to employ them to detect explosives and other important chemical compounds that our weak olfactory senses cannot dream of discerning. Insects are chemical factories, and we are only beginning to discover the possible medical applications of their defensive compounds. Obviously, insects produce raw materials that we use, such as silk, beeswax, honey, shellac, and dyes. Those insects for which we find no utilitarian use are still inherently valuable. They provide ecosystem services such as organic waste removal, pollination of plants, seed dispersal, food for other animals, and pruning of trees and shrubs. We need to conserve the biodiversity of insects if only because we know so little about them and their broader roles in the biosphere.

Those insects that undermine our success we deem to be "pests", despite the fact that such enemies are usually ones that we manufacture. Our worst illegal immigrants are spineless creatures we have imported either intentionally or accidentally and let loose on a new landscape, where they wreak havoc. We plant acres and acres of corn and then complain when the Corn Earworm and European Corn Borer and other insects indulge in their favorite (sometimes only) host plant. Our addiction to chemical insecticides has done more harm than good, even creating new pests by suppressing competitors. We sprayed the boll weevil into submission, only to have the Tobacco Budworm take its place. Because plants defend themselves with toxic chemicals they produce, our synthetic pesticides simply play into the ability of plant-eating insects to eventually adapt to, or detoxify, those compounds. Beneficial predatory and parasitic insects suffer more than the intended targets of our arsenals.

We do not get to choose which insects we share the Earth with, or even our own homes. Our disdain for cockroaches stems from the fact that they exploit our failure to keep a completely crumb-free house or apartment. The fact that bed bugs and human lice can live nowhere *but* on us or with us demotes us unflatteringly to the bottom of the food chain, regardless of our lifestyle and virtue. Ah, but we are not messy housekeepers or unwashed indigents. We are doing our part to promote biodiversity.

These are exciting and troubling times to be following the world of nature and insects in particular. Headlines warn of an "insect Armageddon", scientists sharing research from Germany and Canada, and anecdotal reports from Australia that all suggest insect populations have fallen drastically over the last quarter century. The apiculture (beekeeping) industry clamors for relief in the face of "Colony Collapse Disorder", an apparent conglomeration of detrimental phenomena that is devastating honey bee populations. Meanwhile, other scientists argue that native pollinators, especially solitary bees, are suffering even more in the face of large-scale agricultural practices that ironically make honey bees the only commercially-viable option for pollinating crops. News and dialogue, along with misinformation and industry propaganda, pass at light speed through social media and other online portals. Science is too often left in the dust; communication of facts and research have become increasingly devalued by a society that sees economic, political, and religious points of view as the sole determiners of our future.

We can decide whether we want to turn an epilogue into an epitaph of our wonderful, lively planet by our own personal actions. What we choose to share with others, both in person and across the internet, speaks volumes about what we value and where we place our priorities. Let us hope we continue to be inspired by "the little things that run the world," as E. O. Wilson calls the world of insects. Let us "make peace with the landlord", as the late Thomas Eisner referred to the (ruling) arthropod class. It is not too late to at least mediate our tumultuous relationship.

THE END

Printed in Great Britain
by Amazon